EXTRA LIVES

FALLOUT

HEADSHOTS

THE UNBEARABLE
LIGHTNESS OF GAMES

THE GRAMMAR OF FUN

EXTRA LIVES

WHY VIDEO GAMES MATTER

TOM BISSELL

LITTLEBIGPROBLEMS

BRAIDED

MASS EFFECTS

FAR CRIES

GRAND THEFTS

PANTHEON BOOKS, NEW YORK

Portions of this work originally appeared in The New Yorker,
Tin House, and Kill Screen.

Library of Congress Cataloging-in-Publication Data

Bissell, Tom.
Extra lives : why video games matter / Tom Bissell.
p. cm.
Includes index.
ISBN 978-0-307-37870-5
1. Video games—History.   2. Video Games—Social aspects.
I. Title.
GV1469.3.B55 2010
794.8—dc22   2009039602

www.pantheonbooks.com

Printed in the United States of America

First Edition

2 4 6 8 9 7 5 3

For my brother, Johno, at whom I first threw a joystick

And for my nieces, Amy and Natalie,
who I hope will throw them at me

LYSIMACHUS: Did you go to 't so young? Were you a
    gamester at five or at seven?
MARINA: Earlier too, sir, if now I be one.

—**WILLIAM SHAKESPEARE**, *Pericles*

I've seen things you people wouldn't believe.

—**ROY BATTY** to **RICK DECKARD** in *Blade Runner*

CONTENTS

AUTHOR'S NOTE

M artin Amis, the author of a fine book about early video games, once said of his predicament as a football fan, "Pointy-headed football-lovers are a beleaguered crew, despised by pointy-heads and football-lovers alike." In this book I risk a similar sort of beleagueredness to explain why I believe video games matter—and why they do not matter more. It grew out of the last three years of my life, during which I spent quite a bit—possibly even *most*—of my time playing video games, marveling at the unique ways they affected me and growing frustrated by the ways they did not. Soon enough, I was taking notes, not yet fully aware that what I had actually begun to do was write this book.

Needless to say, it is no easy thing to make a living as a critic of anything, but video-game criticism may be the least remunerative of all. Why this should be is not a great mystery. Count off the number of people of your acquaintance inclined to read criticism at all; chances are lean they will be the same people in your life as the ones playing video games. Yet certain aspects of video games make them resistant to a traditional critical approach. One is that many games are not easily re-experienceable, at least not in the way other mediums are re-experienceable. If I am reviewing a book, I go back and look at my margin notes. An album, I set aside an hour and listen to it again. A film, I buy another ticket. If I am playing a game that takes dozens of hours to complete and has a limited number of save slots, much of it is accessible only by playing it through again, the game itself structurally obligated to fight me every inch of the way. Another problem is that criticism needs a readily available way to connect to the aesthetic past of the form under appraisal, which is not always so easy with video games. Out-of-date hardware and out-of-print games can be immensely difficult to find. Say you want to check on something that happens about halfway through some older game. Not only do you have to find it, you will, once again, have to play it. Probably for hours. Possibly for days.

One might argue that critical writing about games is difficult because most games are not able to withstand thoughtful criticism. For their part, game magazines publish game review after game review, some of which are spritely and sharp, but they tend to focus on providing consumers with a sense of whether their money will be well spent. Game magazine reviewers rarely ask: *What aesthetic tradition does this game fall into? How does it make me feel while I'm playing it? What emotions does it engage with, and are they appropriate to the game's theme and mechanics?* The reason game magazine reviewers do not ask these questions is almost certainly because game magazine owners would like to stay in business. But there *is* a lot of thoughtful, critically engaging work being done on games. It is mostly found on the blogs and almost always done for free. I have my list of the five game critics whose thoughts on the form I am most compelled by, and I am fairly certain that none of these writers is able to make anything resembling a living writing only about games. Certainly, this is the case for the top critics of plenty of other art forms—dance, sculpture, poetry—but none of these art forms is as omnipresent, widely consumed, or profitable as video games.

I say all this up front to signal my awareness that I am far from the first to arrive at this particular party. As a work of criticism, however, this book is somewhat eccentric and, at times, starkly personal. Moreover, its focus falls heavily upon console games (as opposed to personal computer games) released in the last few years, most of which are amply budgeted "story" or "narrative" games, which may displease some readers. From this, no one should assume I am not fond of older games or that I do not play sports games, rhythm games, strategy games, puzzlers, or the like. I am and I do, and moreover will take on any comers willing to challenge me to expert-level drumming in *Rock Band* or *Guitar Hero* (unless you happen to go by the gamertag Johny Red Pants, in which case, I bow to you, fair sir). The fact is, most of the games that made me want to write this book are console games of relatively recent issue, as opposed to the classics of the form. Few mediums are as prone to the evolutionary long jump as the video game, and I am aware that my focus on contemporary games puts these pages in danger of seeming, in only a few years, as relevant as a biology textbook devoted to Lamarckism. While the games I examine may be contemporary and somewhat of a piece, the questions they raise about the video-game form are not likely to lose their relevance anytime soon.

There are many fine books about the game industry, the theory of game design, and the history of games, overmuch discussion of which will not be found here. I did not write this book as an analyst of industry fortunes (a topic about which I could not imagine caring less) or as a chronicler of how games rose and came to be, and my understanding of the technical side of game design is nil. I wrote this book as a writer who plays a lot of games, and in these pages you will find one man's opinions and thoughts on what playing games feels like, why he plays them, and the questions they make him think about. In the portions of the book where I address game design and game designers, it is, I hope, to a formally explanatory rather than technically informative end.

Just what *is* a video game? Decades into the development of the form, this question remains forbiddingly open. (As does the term's spelling: *video game* or *videogame*? I reluctantly prefer the former. Most game designers and critics favor the latter.) It may be years before anyone arrives at a true understanding of what games are, what they have done to popular entertainment, and how they have shaped the wider expectations of their many and increasingly divergent audiences. In my conversations with game designers, I was sternly informed, again and again, of the *newness* of their form, the things they were still learning how to do, and of the necessity of discarding any notion of what defines video games. I have come to believe that anyone who can tell you what a game is, or must be, has seen advocacy outstrip patience. One game designer told me that, due to the impermanent and tech-dependent nature of his medium, he sometimes felt as though he were writing his legacy in water. I nevertheless believe that we are in a golden age of gaming and hope this book will allow future gamers a sense of connection to this glorious, frustrating time, whatever path games ultimately take and whatever cultural fate awaits them.   —TCB

June 1, 2009
Escanaba, Michigan

EXTRA LIVES

FALLOUT

ONE

Someday my children will ask me where I was and what I was doing when the United States elected its first black president. I could tell my children—who are entirely hypothetical; call them Kermit and Hussein—that I was home at the time and, like hundreds of millions of other Americans, watching television. This would be a politician's answer, which is to say, factual but inaccurate in every important detail. Because Kermit and Hussein deserve an honestly itemized answer, I will tell them that, on November 4, 2008, their father was living in Tallinn, Estonia, where the American Election Day's waning hours were a cold, salmon-skied November 5 morning. My intention that day was to watch CNN International until the race was called. I will then be forced to tell Kermit and Hussein about what else happened on November 4, 2008.

The postapocalyptic video game *Fallout 3* had been officially released to the European market on October 30, but in Estonia it was nowhere to be found. For several weeks, Bethesda Softworks, *Fallout 3*'s developer, had been posting online a series of promotional gameplay videos, which I had been watching and rewatching with fetish-porn avidity. I left word with Tallinn's best game store: *Call me the moment* Fallout 3 *arrives.* In the late afternoon of November 4, they finally rang. When I slipped the game into the tray of my Xbox 360, the first polls were due to close in America in two hours. One hour of *Fallout 3*, I told myself. Maybe two. Absolutely no more than three. Seven hours later, blinking and dazed, I turned off my Xbox 360, checked in with CNN, and discovered that the acceptance speech had already been given.

And so, my beloved Kermit, my dear little Hussein, at the moment America changed forever, your father was wandering an ICBM-denuded wasteland, nervously monitoring his radiation level, armed only with a baseball bat, a 10mm pistol, and six rounds of ammunition, in search of a vicious gang of mohawked marauders who were 100 percent bad news and totally had to be dealt with. Trust Daddy on this one.

Fallout 3 was Bethesda's first release since 2006's *The Elder Scrolls IV: Oblivion.* Both games fall within a genre known by various names: the open-world or sandbox or free-roaming game. This genre is superintended by a few general conventions, which include the sensation of being inside a large and disinterestedly functioning world, a main story line that can be abandoned for subordinate story lines (or for no purpose at all), large numbers of supporting characters with whom meaningful interaction is possible, and the ability to customize (or pimp, in the parlance of our time) the game's player-controlled central character. The pleasures of the open-world game are ample, complicated, and intensely private; their potency is difficult to explain, sort of like religion, of which these games become, for many, an aspartame form. Because of the freedom they grant gamers, the narrative- and mission-generating manner in which they reward exploration, and their convincing illusion of endlessness, the best open-world games tend to become leisure-time-eating viruses. As incomprehensible as it may seem, I have somehow spent more than two hundred hours playing *Oblivion*. I know this because the game keeps a running tally of the total time one has spent with it. I can think of only one personal activity I would be less eager to see audited in this way, and it, too, is a single-player experience.

It is difficult to describe *Oblivion* without atavistic fears of being savaged by the same jean-jacketed dullards who in 1985 threw my *Advanced Dungeon & Dragons Monster Manual II* into Lake Michigan. (That I did not even play D&D, and only had the book because I liked to look at the pictures, left my assailants unmoved.) As to what *Oblivion* is about, I note the involvement of orcs and a "summon skeleton" spell and leave it at that. So: two hundred hours playing *Oblivion*? How is that even possible? I am not actually sure. Completing the game's narrative missions took a fraction of that time, but in the world of *Oblivion* you can also pick flowers, explore caves, dive for treasure, buy houses, bet on gladiatorial arena fights, hunt bear, and read books. *Oblivion* is less a game than a world that best rewards full citizenship, and for a while I lived there and claimed it. At the time I was residing in Rome on a highly coveted literary fellowship, surrounded by interesting and brilliant people, and quite naturally mired in a lagoon of depression more dreadfully lush than any before or since. I would be lying if I said *Oblivion* did not, in some ways, aggravate my depression, but it also gave me something with which to fill my days other than piranhic self-hatred. It was an extra life; I am grateful to have had it.

When Bethesda announced that it had purchased the rights to develop *Fallout 3* from the defunct studio Interplay, the creators of the first two *Fallout* games, many were doubtful. How would the elvish imaginations behind *Oblivion* manage with the rather different milieu of an annihilated twenty-third-century America? The first *Fallout* games, which were exclusive to the personal computer, were celebrated for their clever satire and often freakishly exaggerated violence. *Oblivion* is about as satirical as a colonoscopy, and the fighting in the game, while not unviolent, is often weirdly inert.

Bethesda released *Fallout 3*'s first gameplay video in the summer of 2008. In it, Todd Howard, the game's producer, guides the player-controlled character into a disorientingly nuked Washington, DC, graced with just enough ravaged familiarities—among which a pummeled Washington Monument stands out—to be powerfully unsettling. Based on these few minutes, *Fallout 3* appeared guaranteed to take its place among the most visually impressive games ever made. When Bethesda posted a video showcasing *Fallout 3*'s in-game combat—a brilliant synthesis of trigger-happy first-person-style shooting and the more deliberative, turn-based tactics of the traditional role-playing video game, wherein you attack, suffer your enemy's counterattack, counterattack yourself, and so on, until one of you is dead—many could not believe the audacity of its cartoon-Peckinpah violence. Much of it was rendered in a slo-mo as disgusting as it was oddly beautiful: skulls exploding into the distinct flotsam of eyeballs, gray matter, and upper vertebrae; limbs liquefying into constellations of red pearls; torsos somersaulting through the air. The consensus was a bonfire of the skepticisms: *Fallout 3* was going to be fucking awesome.

Needless to say, the first seven hours I spent with the game were distinguished by a bounty of salutary things. Foremost among them was how the world of *Fallout 3* looked. The art direction in a good number of contemporary big-budget video games has the cheerful parasitism of a tribute band. Visual inspirations are perilously few: Forests will be Tolkienishly enchanted; futuristic industrial zones will be mazes of predictably grated metal catwalks; gunfights will erupt amid rubble- and car-strewn boulevards on loan from a thousand war-movie sieges. Once video games shed their distinctive vector-graphic and primary-color 8-bit origins, a commercially ascendant subset of game slowly but surely matured into what might well be the most visually derivative popular art form in history. *Fallout 3* is the rare big-budget game to begin rather than end with its derivativeness.

It opens in 2277, two centuries after a nuclear conflagration between the United States and China. Chronologically speaking, the world this Sino-American war destroyed was of late-twenty-first-century vintage, and yet its ruins are those of the gee-whiz futurism popular during the Cold War. *Fallout 3*'s Slinky-armed sentry Protectrons, for instance, are knowing plagiarisms of *Forbidden Planet*'s Robby the Robot, and the game's many specimens of faded prewar advertising mimic the nascent slickness of 1950s-era graphic design. *Fallout 3* bravely takes as its aesthetic foundation a future that is both six decades old and one of the least convincing ever conceptualized. The result is a fascinating past-future never-never-land weirdness that infects the game's every corner: *George Jetson Beyond Thunderdome.*

What also impressed me about *Fallout 3* was the buffet of choices set out by its early stages. The first settlement one happens upon, Megaton, has been built around an undetonated nuclear warhead, which a strange religious cult native to the town actually worships. Megaton can serve as base of operations or be wiped off the face of the map shortly after one's arrival there by detonating its nuke in exchange for a handsome payment. I spent quite a while poking around Megaton and getting to know its many citizens. What this means is that the first several hours I spent inside *Fallout 3* were, in essence, optional. Even for an open-world game, this suggests an awesome range of narrative variability. (Eventually, of course, I made the time to go back and nuke the place.)

Fallout 3, finally, looks beautiful. Most modern games—even shitty ones—look beautiful. Taking note of this is akin to telling the chef of a Michelin-starred restaurant that the tablecloths were lovely. Nonetheless, at one point in *Fallout 3* I was running up the stairs of what used to be the Dupont Circle Metro station and, as I turned to bash in the brainpan of a radioactive ghoul, noticed the playful, lifelike way in which the high-noon sunlight streaked along the grain of my sledgehammer's wooden handle. During such moments, it is hard not to be startled—even moved—by the care poured into the game's smallest atmospheric details.

Despite all this, I had problems with *Fallout 3,* and a number of these problems seem to me emblematic of the intersection at which games in general currently find themselves stalled. Take, for instance, *Fallout 3's* tutorial. One feels for game designers: It would be hard to imagine a formal convention more inherently bizarre than the video-game tutorial. Imagine that, every time you open a novel, you are forced to suffer through a chapter in which the characters do nothing but talk to one another about the physical mechanics of how one goes about reading a book. Unfortunately, game designers do not really have a choice. Controller schemas change, sometimes drastically, from game to game, and designers cannot simply banish a game's relevant instructions to a directional booklet: That would be a violation of the interactive pact between game and gamer. Many games thus have to come up with a narratively plausible way in which one's controlled character engages in activity comprehensive enough to be instructive but not so intense as to involve a lot of failure. Games with a strong element of combat almost always solve this dilemma by opening with some sort of indifferently conceived boot-camp exercise or training round.

Fallout 3's tutorial opens, rather more ambitiously, with your character's birth, during which you pick your race and gender (if given the choice, I always opt for a woman, for whatever reason) and design your eventual appearance (probably this is the reason). The character who pulls you from your mother's birth canal is your father, whose voice is provided by Liam Neeson. (Many games attempt to class themselves up with early appearances by accomplished actors; Patrick Stewart's platinum larynx served this purpose in *Oblivion.*) Now, aspects of *Fallout 3*'s tutorial are brilliant: When you learn to walk as a baby, you are actually learning how to move within the game; you decide whether you want your character to be primarily strong, intelligent, or charismatic by reading a children's book; and, when the tutorial flashes forward to your tenth birthday party, you learn to fire weapons when you receive a BB gun as a gift. The tutorial flashes forward again, this time to a high school classroom, where you further define your character by answering ten aptitude-test-style questions. What is interesting about this is that it allows you to customize your character *indirectly* rather than directly, and many of the questions (one asks what you would do if your grandmother ordered you to kill someone) are morbidly amusing. While using an in-game aptitude test as a character-design aid is not exactly a new innovation, *Fallout 3* provides the most streamlined, narratively economical, and interactively inventive go at it yet.

By the time I was taking this aptitude test, however, I was a dissident citizen of Vault 101, the isolated underground society in which *Fallout 3* proper begins. My revolt was directed at a few things. The first was *Fallout 3*'s dialogue, some of it so appalling ("Oh, James, we did it. A daughter. Our beautiful daughter") as to make Stephanie Meyer look like Ibsen. The second was *Fallout 3*'s addiction to trust-shattering storytelling redundancy, such as when your father announces, "I can't believe you're already ten," at what is clearly established as your tenth birthday party. The third, and least forgivable, was *Fallout 3*'s Jell-O-mold characterization: In the game's first ten minutes you exchange gossip with the spunky best friend, cower beneath the megalomaniacal leader, and gain the trust of the goodhearted cop. Vault 101 even has a resident cadre of hoodlums, the Tunnel Snakes, whose capo resembles a malevolent Fonz. Even with its backdrop of realized Cold War futurism, a greaser-style youth gang in an underground vault society in the year 2277 is the working definition of a dumb idea. During the tutorial's final sequence, the Tunnel Snakes' leader, your tormentor since childhood, requests your help in saving his mother from radioactive cockroaches (long story), a reversal of such tofu drama that, in my annoyance, I killed him, his mother, and then everyone else I could find in Vault 101, with the most perversely satisfying weapon I had on hand: a baseball bat. Allowing your decisions to establish for your character an in-game identity as a skull-crushing monster, a saint of patience, or some mixture thereof is another attractive feature of *Fallout 3*. These pretensions to morality, though, suddenly bored me, because they were occurring in a universe that had been designed by geniuses and written by Ed Wood Jr.

Had I really waited a year for this? And was I really missing a cardinal event in American history to keep playing it? I had, and I was, and I could not really explain why. I then thought back to those two hundred hours I had spent playing *Oblivion*, a game that had all the afflictions of *Fallout 3* and then some. *Oblivion*'s story has several scenes that are so dramatically overwrought that, upon witnessing one of them, the woman I then lived with announced that she was revoking all vagina privileges until further notice. A friend of mine, another *Oblivion* addict, confessed to playing the game with the volume turned down after his novelist wife's acid dinner-party dismissal of the time he spent "with elves talking bullshit."

What embarrassed me about *Oblivion* was not the elves; it was the bullshit. Similarly, I was not expecting from *Fallout 3* novelistic storytelling and characterization and I was absolutely not expecting realist plausibility. I happily accept that, in the world of *Fallout 3*, heavily armed Super Mutants prowl the streets, two-hundred-year-old rifles remain functional, and your character can recover from stepping in front of a Gatling gun at full bore by drinking water or taking a nap. All of which is obviously preposterous, but *Fallout 3* plays so smoothly that you do not even want to notice. Anyone who plays video games knows that well-designed gameplay is a craft as surely as storytelling is a craft. When gameplay fails, we know it because it does not, somehow, feel *right*. Failed storytelling is more abject. You feel lots of things—just not anything the storyteller wants you to feel.

What I know is this: If I were reading a book or watching a film that, every ten minutes, had me gulping a gallon of aesthetic Pepto, I would stop reading or watching. Games, for some reason, do not have this problem. Or rather, their problem is not having this problem. I routinely tolerate in games crudities I would never tolerate in any other form of art or entertainment. For a long time my rationalization was that, provided a game was fun to play, certain failures could be overlooked. I came to accept that games were generally incompetent with almost every aspect of what I would call traditional narrative. In the last few years, however, a dilemma has become obvious. Games have grown immensely sophisticated in any number of ways while at the same time remaining stubbornly attached to aspects of traditional narrative for which they have shown little feeling. Too many games insist on telling stories in a manner in which some facility with plot and character is fundamental to—and often even determinative of— successful storytelling.

The counterargument to all this is that games such as *Fallout 3* are more about the world in which the game takes place than the story concocted to govern one's progress through it. It is a fair point, especially given how beautifully devastated and hypnotically lonely the world of *Fallout 3* is. But if the world is paramount, why bother with a story at all? Why not simply cut the ribbon on the invented world and let gamers explore it? The answer is that such a game would probably not be very involving. Traps, after all, need bait. In a narrative game, story and world combine to create an experience. As the game designer Jesse Schell writes in *The Art of Game Design,* "The game is not the experience. The game enables the experience, but it is *not the experience.*" In a world as large as that of *Fallout 3,* which allows for an experience framed in terms of wandering and lonesomeness, story provides, if nothing else, badly needed direction and purpose. Unless some narrative game comes along that radically changes gamer expectation, stories, with or without Super Mutants, will continue to be what many games will use to harness their uniquely extravagant brand of fictional absorption.

I say this in full disclosure: The games that interest me the most are the games that choose to tell stories. Yes, video games have always told some form of story. PLUMBER'S GIRLFRIEND CAPTURED BY APE! is a story, but it is a rudimentary fairytale story without any of the proper fairytale's evocative nuances and dreads. Games are often compared to films, which would seem to make sense, given their many apparent similarities (both are scored, both have actors, both are cinematographical, and so on). Upon close inspection the comparison falls leprously apart. In terms of storytelling, they could not be more different. Films favor a compressed type of storytelling and are able to do this because they have someone deciding where to point the camera. Games, on the other hand, contain more than most gamers can ever hope to see, and the person deciding where to point the camera is, in many cases, you—and you might never even see the "best part." The best part of looking up at a night sky, after all, is not any one star but the infinite possibility of what is between stars. Games often provide an approximation of this feeling, with the difference that you can find out what is out there. Teeming with secrets, hidden areas, and surprises that may pounce only on the second or third (or fourth) play-through—I still laugh to think of the time I made it to an isolated, hard-to-find corner of *Fallout 3*'s Wasteland and was greeted by the words FUCK YOU spray-painted on a rock—video games favor a form of storytelling that is, in many ways, completely unprecedented. The conventions of this form of storytelling are only a few decades old and were created in a formal vacuum by men and women who still walk among us. There are not many mediums whose Dantes and Homers one can ring up and talk to. With games, one can.

I am uninterested in whether games are better or worse than movies or novels or any other form of entertainment. More interesting to me is what games *can* do and how they make me feel while they are doing it. Comparing games to other forms of entertainment only serves as a reminder of what games are not. Storytelling, however, does not belong to film any more than it belongs to the novel. Film, novels, and video games are separate economies in which storytelling is the currency. The problem is that video-game storytelling, across a wide spectrum of games, too often feels counterfeit, and it is easy to tire of laundering the bills.

It should be said that *Fallout 3* gets much better as you play through it. A few of its set pieces (such as stealing the Declaration of Independence from a ruined National Archives, which is protected by a bewigged robot programmed to believe itself to be Button Gwinnett, the Declaration's second signatory) are as gripping as any fiction I have come across. But it cannot be a coincidence that every scene involving human emotion (confronting a mind-wiped android who believes he is human, watching as a character close to you suffocates and dies) is at best unaffecting and at worst risible. Can it really be a surprise that deeper human motivations remain beyond the reach of something that regards character as the assignation of numerical values to hypothetical abilities and characteristics?

Viewed as a whole, *Fallout 3* is a game of profound stylishness, sophistication, and intelligence—so much so that every example of Etch A Sketch characterization, every stone-shoed narrative pivot, pains me. When we say a game is sophisticated, are we grading on a distressingly steep curve? Or do we need a new curve altogether? Might we really mean that the game in question only occasionally insults one's intelligence? Or is this kind of intelligence, at least when it comes to playing games, beside the point? How is it, finally, that I keep returning to a form of entertainment that I find so uniquely frustrating? To what part of me do games speak, and on which frequency?

FALLOUT

HEADSHOTS

THE UNBEARABLE
LIGHTNESS OF GAMES

THE GRAMMAR OF FUN

LITTLEBIGPROBLEMS

BRAIDED

MASS EFFECTS

FAR CRIES

GRAND THEFTS

TWO

So it begins here, in your stepfather's darkened living room, with you hunched over, watching as a dateline title card—1998 JULY—forcefully types itself across the television screen. "1998 July"? Why not "England, London"? Why not, "A time once upon"? A narrator debuts to describe something called Alpha Team's in medias res search for something called Bravo Team's downed chopper in what is mouthfully described as a "forest zone situated in the northwest of Raccoon City." Okay. This is a Japanese game. That probably explains the year–date swappage. That also makes "Raccoon City" a valiant attempt at something idiomatically American-sounding, though it is about as convincing as an American-made game set in the Japanese metropolis of Port Sushi. You harbor affection for the products of Japan, from its cuisine to its girls to its video games—the medium Japanese game designers have made their own. To your mind, then, a certain amount of ineffable Nipponese weirdity is par for the course, even if the course in question has fifteen holes and every one is a par nine.

A live-action scene commences in which Alpha Team lands upon a foggy moor, finds Bravo Team's crashed chopper, and is attacked by Baskervillian hounds, but all you are privy to is the puppetry of snarling muzzles shot in artless close-up. To the canine puppeteers' credit, the hounds are more convincing than the living actors, whose performances are miraculously unsuccessful. The cinematography, meanwhile, is a shaky-cam, *Evil Dead*–ish fugue minus any insinuation of talent, style, or coherence. Once the hellhound enfilade has taken the life of one Alpha Team member, the survivors retreat into a nearby mansion. You know that one of these survivors, following the load screen, will be yours to control. Given the majestic incompetence of the proceedings thus far, you check to see that the game's receipt remains extant.

For most of your life you have played video games. You have owned, in turn, the Atari 2600, the Nintendo Entertainment System, the Sega Genesis, the Super Nintendo, and the Nintendo 64, and familiarized yourself with most of their marquee titles. The console you are playing now, the console you have only today purchased, is categorically different from its ancestors. It is called the Sony PlayStation. Its controllers are more ergonomic than those you have previously held and far more loaded with buttons, and its games are not plastic cartridges but compact discs. Previous consoles were silent but your new PlayStation zizzes and whirs in an unfamiliar way as its digital stylus scans and loads.

It is 1997. The PlayStation was released to the American market one year ago. You missed this, having been away, in the Peace Corps, teaching English, which service you terminated in a panic sixteen months short of your expected stay. Now you are back in your hometown, in the house you grew up in, feeling less directionless than mapless, compassless, in lack of any navigational tool at all. You are also bored. Hence the PlayStation.

The live-action sequence has given way to an animated indoor tableau of surprising detail and stark loveliness—like no console game you have hitherto encountered. Three characters stand in the mansion foyer. There is Barry, a husky, ursine, ginger-bearded man; Wesker, enjoying the sunglasses and slicked-back hair of a coke fiend; and Jill, your character, a trim brunette looker in a beret. A brief conversation ensues about the necessity of finding Chris, your fellow Alpha Team member, who has somehow managed to go AWOL in the time it took to step across the threshold of the mansion's entryway. Soon enough, a gunshot sounds from the next room. You and Barry are dispatched by Wesker to investigate.

The dialogue, bad enough as written ("Wow. What a mansion!"), is mesmerizing in performance. It is as though the actors have been encouraged to place emphasis on the least apposite word in every spoken line. Barry's "He's our old partner, you know," to provide but one example, could have been read in any number of more or less appropriate ways, from "*He's* our old partner, you know" to "He's *our* old partner, you know" to "He's our old *partner,* you know." "He's our *old* partner, you *know*" is the line reading of autistic miscalculation this game goes with.

Upon entry into the new room, you are finally granted control of Jill, but how the game has chosen to frame the mise-en-scène is a little strange. You are not looking through Jill's eyes, and movement does not result in a scrolling, follow-along screen. Instead Jill stands in what appears to be a dining room, the in-game camera angled upon her in a way that annuls any wider field of vision. Plenty of games have given you spaces around which to wander, but they always took care to provide you with a maximal vantage point. This is not a maximal angle; this is not at all how your eye has been trained by video games to work. It as though you, the gamer, are an invisible, purposefully compromised presence within the gameworld.

The room's only sound is a metronomically ticking grandfather clock. You step forward, experimenting with your controller's (seventeen!) buttons and noting the responsiveness of the controls, which lend Jill's movement a precision that is both impressive and a little creepy. Holding down one button allows Jill to run, for instance, and this is nicely animated. A pair of trigger buttons lie beneath each of your index fingers. Squeeze the left trigger and Jill lifts her pistol into firing position. Squeeze the right trigger and Jill fires, loudly, her pistol kicking up in response. All of this—from the preparatory prefiring mechanic to the unfamiliar sensation of consequence your single shot has been given—feels new to you. Every video-game gun you have previously fired did so at the push of a single button, the resultant physics no more palpable or significant than jumping or moving or any other in-game movement. Video-game armaments have always seemed to you a kind of voodoo. If you wanted some digital effigy to die, you simply lined it up and pushed in the requisite photonic pin. Here, however, there is no crosshair or reticule. You fire several more shots to verify this. How on earth do you aim?

As you explore the dining room something even more bizarre begins to occur. The in-game camera is *changing angles*. Depending on where you go, the camera sometimes frames your character in relative close-up and, other times, leaps back, reducing Jill to an apparent foreground afterthought. And yet no matter the angle from which you view Jill, the directional control schema, the precision of which you moments ago admired, remains the same. What this means is that, with every camera shift, your brain is forced to make a slight but bothersome spatial adjustment. The awkwardness of this baffles you. When you wanted Link or Mario to go left, you pushed left. That the character you controlled moved in accordance to his on-screen positioning, which in turn corresponded to your joystick or directional pad, was an accepted convention of the form. Yes, you have experienced "mode shifts" in games before—that, too, is a convention—but never so inexplicably or so totally. So far, the game provides no compelling explanation as to why it has sundered every convention it comes across.

The dining room itself is stunning, though, reminding you of the flat lush realism of *Myst,* a personal computer game your girlfriend adores but that has always struck you as a warm-milk soporific. You have not played a tremendous number of PC games; it is simply not a style of gaming you respond to. You are a console gamer, for better or worse, even though you are aware of the generally higher quality of PC games. Anyone who claims allegiance to the recognizably inferior is in dire need of a compelling argument. Here is yours: The keyboard has one supreme purpose, and that is to create words. Swapping out keys for aspects of game control (J for "jump," < for "switch weapon") strikes you as frustrating and unwieldy, and almost every PC game does this or something like it. PC gamers themselves, meanwhile, have always seemed to you an unlikable fusion of tech geek and cult member—a kind of mad Scientologist.

You glance at the box in which this game came packaged. *Resident Evil.* What the hell does that even *mean*? You know this game is intended to be scary. You also know that zombies are somehow involved; the box art promises that much. The notion of a "scary game" is striking you as increasingly laughable. While nothing is more terrifying to you than zombies, calling a zombie-based game *Resident Evil* is a solecism probably born of failing to fully understand the zombie. Part of what makes zombies so frightening is that they are *not* evil. The zombie, a Caribbean borrowing, is in its North American guise a modern parable for . . . well, there you go. Like all parables, zombies are both widely evocative and impossible to pin down. Part of the reason you purchased this game was because you were curious to see what the Japanese imagination had made of the zombie. This was a culture, after all, that had transformed its twentieth-century resident evil into a giant bipedal dinosaur.

On screen, Barry calls Jill over, where he kneels next to a pool of blood. ("I hope it's not . . . *Chris's* blood.") He orders you to press on looking around while he completes his investigation. You are no criminologist, but gleaning the available information from a small, freestanding blood puddle would seem to you an undertaking of no more than three or four seconds. Barry, though, continues to ponder the hell out of that blood. You have two options. Leave the dining room to go back and explore the foyer, where Wesker presumably awaits your report, or go through a nearby side door. You take the side door. Anytime you go through a door in this game you are presented with a load screen of daunting literalness: the point of view reverts to an implied first-person, the door grows closer, the knob turns, the door opens, which is followed by the noise of it closing behind you. Considerable investment has been placed in a dramatic reproduction of this process: The knobs sound as though they were last oiled in the Cleveland administration, and the doors themselves slam shut as though they weigh five hundred pounds.

The load screen complete, Jill now stands in a long narrow hallway. The camera looks down upon her from an angle of perhaps seventy degrees, which leaves you unable to see either ahead of or behind her. You turn her left, instinctively, only to hear something farther down the hall. You hear . . . *chewing*? No. It is worse than that. It is a wet, slushy sound, more like *feasting* than chewing. The camera has shifted yet again, allowing you to look down the hall but not around the corner, whence this gluttonous feasting sound originates. There is no music, no cues at all. The game-world is silent but for your footsteps and the sound you now realize you have been set upon this path to encounter. You panic and run down to the other end of the hall, the feasting sound growing fainter, only to find two locked doors. No choice, then. You walk (not run) back toward the hallway corner, then stop and go to a subscreen to check your inventory. Your pistol's ammunition reserves are paltry, and you curse yourself for having shot off so many bullets in the dining room. You also have a knife. You toggle back and forth between pistol and knife, equipping and unequipping. You eventually go with the pistol and leave the inventory screen.

Jill stands inches before the hallway corner, but it suddenly feels as though it is you standing before hellmouth itself. Your body has become a hatchery from which spiderlings of dread erupt and skitter. Part of this is merely expectation, for you know that a zombie is around that corner and you are fairly certain it is eating Chris. Another part is . . . you are not sure you can name it. It is not quite the control-and-release tension of the horror film and it is not quite actual terror. It is something else, a fear you can control, to a point, but to which you are also helplessly subject— a fear whose electricity becomes pleasure.

You raise your pistol—and this is interesting: You *cannot move* while your pistol is raised. You had not noticed this before. You should be able to move with your pistol raised, and certainly you should be able to shoot while moving. That is another convention of the form. In video games, you can shoot your sluggish bullets while running, jumping, falling off a cliff, swimming underwater. On top of this you have exactly five rounds. Zombies are dispatched with headshots. You know that much. But how do you shoot for the head when the game provides you with no crosshair? A "scary game" seems a far less laughable notion than it did only a few moments ago.

You turn the corner to yet another camera change. You have only a second or two to make out the particulars—a tiny room, a downed figure, another figure bent over him—before what is called a cut scene kicks in. The camera closes on a bald humanoid, now turning, noticing you, white head lividly veiny, mouth bloody, eyes flat and empty and purgatorial. There the brief cut scene ends. The zombie, now approaching, groans in thoughtless zombie misery, a half-eaten corpse behind it. You fire but nothing happens. In your panic you have forgotten the left trigger, which raises your weapon. This blunder has cost you. The zombie falls upon you with a groan and bites you avidly, your torso transforming into a blood fountain. You mash all seventeen of your controller's buttons before finally breaking free. The zombie staggers back a few steps, and you manage to fire. Still no crosshair or reticule. Your shot misses, though by how much you have no idea. The zombie is upon you again. After pushing it away—and there is something date-rapeishly unwholesome about the way it assaults you—you stagger back into the hallway to give yourself more room to maneuver, but the camera switches in such a way as to leave you unaware of the zombie's exact location, though you can still hear its awful, blood-freezing moan, which, disembodied, sounds not only terrifying but *sad*. You fire blindly down the hall, toward the moaning, with no guarantee that your shots are hitting the zombie or coming anywhere close to it. Soon pulling the trigger produces only spent clicks. You go to the inventory screen and equip your knife. When you return to gameplay, the zombie appears within frame and lurches forward. You slash at it, successfully, blood geysering everywhere, but not before it manages to grab on to you yet again. After another chewy struggle, you back up farther, the camera finally providing you with a vantage point that is not actively frustrating, and you lure the zombie toward you, lunging when it staggers into stabbing range. At last the creature drops. You approach its doubly lifeless husk, not quite believing what is happening when it grabs your leg and begins, quite naturally by this point, to bite you. You stab at this specimen of undead indestructibility until, with a final anguished moan, a copious amount of blood pools beneath it. What new devilry is this?

None of it has made sense. Not the absurd paucity of your ammunition stores, not the handicapping camera system, not the amount of effort it took to defeat a single foe, not that foe's ability to play dead. You know a few things about video-game enemies. When they are attacked they either die instantly or lose health, and for foes as tough as this one you are typically able to track the process by way of an onscreen health bar. This zombie, however, had no health bar. (Neither do you, properly speaking. What you do have is an electrocardiographic waveform that is green when you are at full health, orange when you are hurt, and red when you are severely hurt. Not only is this EKG stashed away in the inventory subscreen, it provides only an approximate state of health. Right now your health is red. But *how* red? You have no idea. This game is rationing not only resources but *information*.) When video-game characters die, furthermore, they disappear, like Raptured Christians or Jedi. Your assailant has not disappeared and instead remains facedown in a red pool of useless zombie plasma. This is a game in which every bullet, evidently, will count. This is also a game in which everything you kill will remain where it falls, at least until you leave the room. You stab it again. Revenge!

You flee the hallway and return to Barry. Before you can tell him what has happened, the door behind you opens. The zombie whose deadness was a heliocentric certainty has followed you. You (not Jill: *you*) cry out in delighted shock. Your worried step-father, a few rooms away, calls your name, his voice emanating from a world that, for the last half hour, has been as enclosing but indistinct as an amnion. After calling back that you are okay, you are newly conscious of the darkness around you, the lateness of the hour. For the first time in your life, a video game has done something more than entertain or distract you. It has bypassed your limbic system and gone straight for the spinal canal. You lean back, cautiously. You are twenty-three years old. You have played a lot of games. Right now, all those games, all the irrecoverable eons you have invested in them, seem to you, suddenly, like nothing more than a collective prologue.

The critic Robert Hughes called it "the shock of the new": the sensation of encountering a creative work that knocks loose the familiar critical vocabularies and makes them feel only partially applicable to what stands before you. It is the powerful, power-less feeling of knowing your aesthetic world has been widened but not yet having any name for the new ground upon which you stand. Hughes was talking about visual art, but there is no reason to confine the shock of the new to any particular medium. When it comes to video games, the shock of the new came to me through Capcom's *Resident Evil,* though other gamers will have their own equally resonant examples. The first time I played *Resident Evil* is the only instance in which I was acutely aware of being present at the birth of a genre (that of "survival horror"), and it was one of a handful of occasions that a medium I believed I understood felt objectively, qualitatively *new*—and not merely new to me.

At first glance *Resident Evil* seemed to be imitating horror films: the grindingly familiar character types; the shifting camera angles;

the elongated, tonal creepiness occasionally punctuated by sudden, decisive scares; the brilliant—absolutely *brilliant*—use of sound. But it also took core inspiration from primitive video-game progenitors. Much of *Resident Evil* involves finding objects (a lighter, herbs of various colors, sheet music, jewelry) and figuring out how to use them and where, a this-quest-opens-that-quest structure similar to some of the earliest text-based computer adventures, one of which was actually called *Adventure*. *Resident Evil*'s reliance on gunplay—it was originally envisioned as a first-person shooter—came from games too numerous to mention. When it was not borrowing horror-film decor, *Resident Evil* frequently resembled, as mentioned, *Myst*. None of these constituent parts was new, but the unlikely whole they formed was. No game had ever before combined so many disparate strands of popular entertainment; few had pointed more evocatively to what was possible within the video-game form.

Oddly, not many games chose to follow where *Resident Evil* pointed. Its innovations were selectively scavenged rather than swallowed whole, even within subsequent *Resident Evil* titles. The intentional clumsiness of the controls (cardiac-event-inducing when surrounded by shambling, moaning zombies) was abandoned by *Resident Evil 4*, as was the cinematically relocating camera. The former was dropped because it was no longer an interesting hindrance; gamers had learned to adapt to it. The latter was dropped because the game's designers settled upon more direct ways to alarm gamers than by the obscurantism of shifting camera angles. In *Resident Evil 4*, they simply throw more enemies at you than you can ever hope to kill.

The innovations that did survive are a mixed bag. Most narrative games today require the player to "save" his or her progress. In early games you were often given a password to allow you to start where you last left off; later games did the saving for you, automatically. *Resident Evil's* save system—which involved, for some deeply mysterious reason, finding in-game typewriter ribbons, which one then used to save one's progress on an in-game old-fashioned manual typewriter—was about as frustrating as typing on an out-of-game old-fashioned manual typewriter. You could save only in special typewriter-having locations, and your severely limited inventory space meant you could only carry so many items at once (fewer still if you picked Jill to control rather than brawny Chris), the upshot of which was spending half the game muttering profanities while running back and forth through rooms filled with undead to "save" rooms and then swapping out items to make space for your latest stumbled-upon typewriter ribbon and then running back to fetch it. Saving your game at every opportunity became an imperative as biologically intense as food or sleep. I have had friends and relatives die, lovers stray, and money run out, but I think I would still place being torn apart by zombies with an hour and a half of unsaved *Resident Evil* gameplay behind me in the upper quartile of Personally Miserable Experiences. While the satanically complicated save system certainly upped the tension of *Resident Evil's* gameplay, it did so artificially, and for years a number of games, especially Japanese games, made saving one's progress a similarly and uselessly time-consuming ordeal. This succeeded grandly in making games harder but did nothing to make them more enjoyable. (Ten years later, another Capcom zombie game, *Dead Rising,* would have an even more infuriating and niggardly save system. As much as I love *Dead Rising,* I still wish ill upon everyone involved with its save-system implementation. Honestly. Those people can go to hell.)

One of *Resident Evil's* more influential innovations did not concern gameplay per se. Games were violent before *Resident Evil,* certainly, but they were violent in two ways: operatically (as in *Mortal Kombat's* "Finish him!" finale scenes) or iteratively (the mow-'em-down mindlessness of just about everything else). The violence of *Resident Evil* was surprisingly occasional but unbelievably brutal. It was also clinical, which encouraged a certain wicked tendency to experiment. As you found and used new weapons, it turned out that zombies reacted to them in varied ways. A shotgun could blow the legs out from under a zombie, and the well-placed round of a .38 could take its head right off. I do not claim to be a historian of video-game dismemberment, but I am fairly sure that no game before *Resident Evil* allowed such violence to be done to specific limbs. It provided gamers with one of the video-game form's first laboratories of virtual sadism, and I would be lying if I did not admit that it was, in its way, exhilarating. (They were *zombies.* You were doing them a *favor.*)

But *Resident Evil* was influential in a final, lamentable way, and this has to do with its phenomenal stupidity. How stupid was *Resident Evil*? So stupid that stupidity has since become one of the signatures of the *Resident Evil* series. So stupid, in other words, that stupidity became something not to address or fix but a mast of tonal distinction to which the series lashed itself. I have already quoted some of the game's dialogue, which at its least weird sounds as though it has been translated out of Japanese, into Swahili, back into Japanese, into the language of the Lunar Federation, back into Japanese, and finally into English. As for the plot, I have played through the game at least half a dozen times and could not under pain of death explain its most rudimentary aspects. I know that the plot provides a stage for the considerable malversation of your erstwhile teammate Wesker. I also know that it involves an evil corporation known as Umbrella and a terrible biotoxin known as the T-virus. This is where the cinematic sweep and texture of *Resident Evil* least resemble cinema. Great horror movies are almost always subterranean in effect. They are the ultimate compulsion—*you must watch*—and they transubstantiate social anxieties more sensed than felt. The sensed, rather than the felt, is the essence of the horror film. Another way of saying this is that good horror films are *about* something not immediately discernible on their surface. On its surface, *Resident Evil* is about an evil corporation known as Umbrella and a terrible biotoxin known as the T-virus. Beneath that surface is a tour de force of thematic nullity. All the game really wants to do is frighten you silly, and it goes about doing so with considerable skill. Playing it for the first time was easily as scary as any horror movie and frequently much scarier. But was it horrifying? For me, horror is the departure of conscious thought, and *Resident Evil* collapses wherever thought arrives.

This brilliantly conceived game of uncompromising stupidity was, in retrospect, a disastrous formal template. Terrible dialogue? It was still a great game. A constant situational ridiculousness that makes *The Texas Chain Saw Massacre* seem like a restrained portrait of rural dysfunction? It was still a great game. And it *is* a great game, and will be ever thus. It was eventually remade, more than once, most notably for release on the Nintendo GameCube, with better graphics and voice actors and a script translated by someone who had occasionally heard spoken English. It, too, was a great game. But the success of the first *Resident Evil* established the *permissibility* of a great game that happened to be stupid. This set the tone for half a decade of savagely unintelligent games and helped to create an unnecessary hostility between the greatness of a game and the sophistication of things such as narrative, dialogue, dramatic motivation, and characterization. In accounting for this state of affairs, many game designers have, over the years, claimed that gamers do not much think about such highfaluting matters. This may or may not be largely true. But most gamers do not care because they have been trained by game designers not to care.

Without a doubt, *Resident Evil* showed how good games could be. Unfortunately, it also showed how bad games could be. Too amazed by the former, gamers neglected to question the latter. It rang a bell to which too many of us still, and stupidly, salivate.

FALLOUT

HEADSHOTS

THE UNBEARABLE LIGHTNESS OF GAMES

THE GRAMMAR OF FUN

LITTLEBIGPROBLEMS

BRAIDED

MASS EFFECTS

FAR CRIES

GRAND THEFTS

THREE

I have been publishing long enough now to look back on much of what I have written and feel the sudden, pressing need to throw myself off the nearest bridge. Every person lucky enough to turn a creative pursuit into a career has these moments, and at least, I sometimes tell myself, I do not often look back on my writing with shame.

I am ashamed of one thing, however, and that is an essay I contributed to a nonfiction anthology of "young writing." I was encouraged to write about anything I pleased, so long as it addressed what being a young writer today felt like. I wrote about video games and whether they were a distraction from the calling of literature. Even as I was writing it, I was aware that the essay did not accurately reflect my feelings. Recently I wondered if the essay was maybe somewhat better than I remembered. I then reread it and spent much of the following afternoon driving around, idly looking for bridges.

"As for video games," I wrote, "very few people over the age of forty would recognize them as even a lower form of art. I am always wavering as to where I would locate video games along art's fairly forgiving sliding scale." Video games are obviously and manifestly a form of popular art, and every form of art, popular or otherwise, has its ghettos, from the crack houses along Michael Bay Avenue to the tubercular prostitutes coughing at the corner of Steele and Patterson. The video game is the youngest and, increasingly, most dominant popular art form of our time. To study the origins of any popular new medium is to become an archaeologist of skeptical opprobrium. It seems to me that anyone passionate about video games has better things to do than walk chin-first into sucker-punch arguments about whether they qualify as art. Those who do not believe video games are or ever will be art deserve nothing more goading or indulgent than a smile.

I think that was what I was trying to say. But I was then and am now routinely torn about whether video games are a worthy way to spend my time and often ask myself why I like them as much as I do, especially when, very often, I hate them. Sometimes I think I hate them because of how purely they bring me back to childhood, when I could only imagine what I would do if I were single-handedly fighting off an alien army or driving down the street in a very fast car while the police try to shoot out my tires or told that I was the ancestral inheritor of some primeval sword and my destiny was to rid the realm of evil. These are very intriguing scenarios if you are twelve years old. They are far less intriguing if you are thirty-five and have a career, friends, a relationship, or children. The problem, however, at least for me, is that they are no less *fun*. I like fighting aliens and I like driving fast cars. Tell me the secret sword is just over the mountain and I will light off into goblin-haunted territory to claim it. For me, video games often restore an unearned, vaguely loathsome form of innocence—an innocence derived of *not knowing anything*. For this and all sorts of other complicated historical reasons—starting with the fact that they began as toys directly marketed to children—video games crash any cocktail-party rationale you attempt to formulate as to why, exactly, you love them. More than any other form of entertainment, video games tend to divide rooms into Us and Them. We are, in effect, admitting that we like to spend our time shooting monsters, and They are, not unreasonably, failing to find the value in that.

I wrote in my essay that art is "obligated to address questions allergic to mere entertainment. . . . In my humble estimation, no video game has yet crossed the Rubicon from entertainment to true art." Here I was trying to say that what distinguishes one work of art from another is primarily intelligence, which is as multivalent as art itself. Artistic or creative intelligence can express itself formally, stylistically, emotionally, thematically, morally, or any number of ways. Works of art we call masterpieces typically run the table on the many forms artistic intelligence can take: They are comprehensively intelligent. This kind of intelligence is most frequently apparent in great works of art created by individuals. Unity of artistic effect is something human beings have learned to respond to, and for obvious reasons this is best achieved by individual artists. Many games—which are, to be sure, corporate entertainments created by dozens of people with a strong expectation of making a lot of money—have more formal and stylistic intelligence than they know what to do with and not even trace amounts of thematic, emotional, or moral intelligence. One could argue that these games succeed as works of art in some ways and either fail or do not attempt to succeed in others. "True" art makes the attempt to succeed in every way available to it. At least, I think so.

My ambivalence goes much deeper, though. A few years ago I was asked by a magazine for my year-end roundup of interesting aesthetic experiences, among which I included 2K Boston's peerless first-person shooter *BioShock,* which, I wrote, "I would hesitate to call . . . a legitimate work of art," even though "its engrossing and intelligent story line made it the first game to absorb me without also embarrassing me for being so absorbed." Seeing that half-hearted encomium in print, with my name attached to it, about a game I adored, obsessed over, and thought about for weeks drove home the plunger of a fresh syringe of shame. Was I apologizing to some imaginary cultural arbiter for finding value in a form of creative expression whose considerable deficits I recognize but which I nevertheless believe is important? Or is this evidence of an authentic scruple? On one hand, I love *BioShock,* which is frequently saluted as one of the first games to tackle what might be considered intellectual subject matter—namely, a gameworld exploration of the social consequences inherent within Ayn Rand's Objectivism (long story). On the other hand, what passes for intellectual subject matter in a video game is still far from intellectually compelling, at least to me, and I know I was not imagining the feeling of slipping, hourglass loss I experienced when I played *BioShock* ten hours a day for three days straight. If I really wanted to explore the implications and consequences of Objectivism, there were better, more sophisticated places to look, even if few of them would be as much fun (though getting shot in the knee would be more fun than rereading *Atlas Shrugged*). When I think about games, here is where I bottom out. Is it okay that they are *mostly* fun? Am I a philistine or simply a coward? Are games the problem, or am I?

I came to this once-embarrassed, formerly furtive love of games honestly. Because the majority of the games I have enjoyed most as an adult tell stories, I was always comparing those stories with the novels and films I admire. Naturally, I found (and find) most video-game stories wanting. But this may be a flagrant category mistake. For one thing, no one is sure what purpose "story" actually serves in video games. Games with any kind of narrative structure usually employ two kinds of storytelling. One is the framed narrative of the game itself, set in the fictional "present" and traditionally doled out in what are called cut scenes or cinematics, which in most cases take control away from the gamer, who is forced to watch the scene unfold. The other, which some game designers and theoreticians refer to as the "ludonarrative," is unscripted and gamer-determined—the "fun" portions of the "played" game—and usually amounts to some frenetic reconception of getting from point A to point B. The differences between the framed narrative and the ludonarrative are what make story in games so unmanageable: One is fixed, the other is fluid, and yet they are intended, however notionally, to work together. Their historical inability to do so may be best described as congressional.

An example of such narrative cross-purpose can be found in Infinity Ward's first-person shooter *Call of Duty 4*. In one memorable sequence, moving forward the framed narrative requires you and a computer-controlled partner to crawl and sneak your way through the irradiated farmlands of Chernobyl in order to assassinate an arms dealer. The ludonarrative, meanwhile, is the actual (and, as it happens, pretty thrilling) process of getting there. If you choose to be a dick and frag your partner, it has only ludonarrative consequences. At worst, you have to start the mission over. No matter what you do, the framed narrative does not change: You and he need to get there together. *Call of Duty 4* is a game with little to no ambition to change the emotional outlook of anyone who plays it. It is a war-porn story of good and evil. All the same, the chasm between its framed narrative and ludonarrative calls attention to the artificiality of both. While the former attempts to be narratively meaningful, the latter is concerned only with being exciting. The former grants the player no agency and thus has no emotional resonance because the latter, with its illusion of agency, does nothing to reinforce what that resonance might be, other than that shooting your friend in the head is bad news. Believing in the game's fiction often becomes as difficult as obeying orders issued by a world-class hypocrite. For a game of *Call of Duty 4*'s simplistic themes, this is a problem of glancing consequence. For games of greater ambition, however, the problem becomes exponentially larger. (*Call of Duty 4* does offer a couple of formally compelling experiences. One is that it kills off the character you assume you will control for the duration in a mid-game helicopter crash, but not before allowing you to take a few disoriented steps from the wreckage—altogether an eerie sequence. Another is the game's opening, which grants the gamer the helpless first-person POV of a man being driven, it becomes increasingly evident, to his execution. This sequence ends with the gamer being shot, jarringly, in the face.)

Several games have lately been experimenting with allowing decisions made during the ludonarrative to alter the framed narrative, most notably in *Fallout 3* and Lionhead's *Fable II*, but this is mainly expressed in how you are perceived by other characters. Once a game comes along that figures out a way around the technical challenges of allowing a large number of ludonarrative decisions to have framed-narrative-altering consequences—none of which challenges I understand but whose existence several game designers sighingly confirmed for me—an altogether new form of storytelling might be born: stories that, with your help, create themselves. There is, of course, another word for stories that, with your help, create themselves. That word is *life.* So would this even be a good thing?

I am not so sure. When I am being entertained, I am also being manipulated. I am *allowing* myself to be manipulated. I am, in other words, surrendering. When I watch television, one of our less exalted forms of popular entertainment, I am surrendering to the inevitability of commercials amid bite-sized narrative blocks.

When I watch a film, the most imperial form of popular entertainment—particularly when experienced in a proper movie
theater—I am surrendering most humiliatingly, for the film begins
at a time I cannot control, has nothing to sell me that I have not
already purchased, and goes on whether or not I happen to be in
my seat. When I read a novel I am not only surrendering; I am
allowing my mind to be occupied by a colonizer of uncertain
intent. Entertainment takes it as a given that I cannot affect it other
than in brutish, exterior ways: turning it off, leaving the theater,
pausing the disc, stuffing in a bookmark, underlining a phrase.
But for those television programs, films, and novels febrile with
self-consciousness, entertainment pretends it is unaware of me,
and I allow it to.

Playing video games is not quite like this. The surrender is
always partial. You get control and are controlled. Games are
patently aware of you and have a physical dimension unlike any
other form of popular entertainment. On top of that, many require
a marathon runner's stamina: Certain console games can take as
many as forty hours to complete, and, unlike books, you cannot
bring them along for enjoyment during mass-transit dead time.
(Rarely has wide-ranging familiarity with a medium so transpar-
ently privileged the un- and underemployed.) Even though you
may be granted lunar influence over a game's narrative tides, the
fact that there is any narrative at all reminds you that a presiding
intelligence exists within the game along with you, and it is this
sensation that invites the otherwise unworkable comparisons
between games and other forms of narrative art.

Yes, as difficult as it sometimes is to believe, games have
authors, however diminutive an aura he or she (or, frequently,
they) might exude. What often strikes me whenever I am playing
a game is how glad I am of that hovering authorial presence.
Although I enjoy the freedom of games, I also appreciate the

remindful crack of the narrative whip—to seek entertainment is to seek that whip—and the mixture of the two is what makes games such a seductive, appealingly dyadic form of entertainment. A video game whose outcomeless narrative is wholly determined by my actions—as in, say, *World of Warcraft,* which is less a video game than a digital board game, and which game I very much dislike—would elevate me into a position of accidental authorship I do not covet and render the game itself a chilly collation of behavior trees and algorithms. I *want* to be told a story—albeit one I happen to be part of and can affect, even if in small ways. If I wanted to *tell* a story, I would not be playing video games.

A noisy group of video-game critics and theoreticians laments the rise of story in games. Games, in one version of this view, are best exemplified as total play, wherein the player is an immaterial demiurge and the only "narrative" is what is anecdotally generated during play. (*Tetris* would be the best example of this sort of game.) My suspicion is that this lament comes less from frustration with story qua story than it does from the narrative butterfingers on outstanding display in the vast majority of contemporary video games. I share that frustration. I also love being the agent of chaos in the video-game world. What I want from games—a control as certain and seamless as the means by which I am being controlled—may be impossible, and I am back to where I began. Reload.

The purpose story serves in video games is complicated, then. Less complicated is how many gamers view story. For many gamers (and, by all evidence, game designers), story is largely a matter of accumulation. The more *explanation* there is, the thought appears to go, the more *story* has been generated. This would be a profound misunderstanding of story for any form of narrative art, but it has hobbled the otherwise high creative achievement of any number of games. Frequently in work with any degree of genre loyalty—this would include the vast majority of video games—the more explicit the story becomes, the more silly it will suddenly seem. (Let us call this the Midi-chlorian Error.) The best science fiction is usually densely realistic in quotidian detail but evocatively vague about the bigger questions. Tolkien is all but ruined for me whenever I make the mistake of perusing the Anglo-Saxon Talmudisms of his various appendices: "Among the Eldar the Alphabet of Daeron did not develop true cursive forms"—kill me, please, now—"since for writing the Elves adopted the Fëanorian letters." As for horror films, the moment I learned Freddy Krueger was "the bastard son of a thousand maniacs" was also the final moment I could envision him without spontaneously laughing. The impulse to *explain* is the Achilles' heel of all genre work, and the most sophisticated artists within every genre know better than to expose their worlds to the sharp knife of intellection.

A good example of a game that does not make that mistake is Valve's cooperative first-person shooter *Left 4 Dead,* which offers yet another vision of zombie apocalypse. Unlike the *Resident Evil* series, which goes to great narrative pains to explain what is happening and why (culminating in one of the most ridiculous moments in video-game history, when the hero of *Resident Evil 4* discovers an enemy document helpfully titled OUR PLAN), *Left 4 Dead* abandons every rational pretext and drops you and three other characters into the middle of undead anarchy. Almost nothing is explained; the little characterization there is comes in tantalizing dribs; and all that is expected is survival, which is possible only by constantly working together with your fellow gamers: covering them while they reload, helping them up when they are knocked down, and saving them when they are trapped in the eye of a zombie hurricane. *Left 4 Dead* is one of the most well-designed and explosively entertaining games ever made. While its purpose is incontinent terror, its point is that teamwork is, by definition, a matter of compulsion, not choice. *Left 4 Dead*'s designer, Michael Booth, had the maturity to grasp the power that narrative minimalism would provide his game. The speedy and acrobatic zombies of *Left 4 Dead* have no plan more refined than kicking you to death and sucking the marrow from your femur. As a scenario, it is as ridiculous as any forged by the Vulcans of video-game conceit, and yet, from start to finish, *Left 4 Dead* is as freefallingly unfamiliar and viscerally convincing as the worst dream you have ever had.

Capturing what playing *Left 4 Dead* feels like is not easy. But set *Left 4 Dead* to its highest difficulty level, recruit three of its best players you can find, push your way through one of the game's four scenarios, and make no mistake: What will go down will be so emotionally grueling, it will feel as though you have spent an hour playing something like full-contact psychic football. The end of the game, however it turns out, will feel epic to no one who did not take part in it, but those who did take part will feel as though they have marched, together, through a gauntlet of the damned.

The game's refusal to explore the who, what, why, or how of its zombie citizenry is emblematic of the unusually austere approach to narrative in many Valve games, which the company may not have invented but has certainly come close to perfecting. The four controllable characters in *Left 4 Dead* are all common video-game types: the girl, the black guy, the biker, the elderly Vietnam vet. They are not, however, blank canvases. (I play as—in order of preference—the girl, the black guy, and the biker. I absolutely refuse to play as the Vietnam vet. For some reason I absolutely hate the guy. Tactics that failed in the jungles and swamps of the Mekong Delta have no place against an army of the undead.) The object of the game is to fight your way through scenarios that are themselves divided into five stages, all of which, but for the scenarios' finales, conclude with the players' slamming shut a safe house's thick red metal door. The problem, of course, is that between these safe houses are devastated locales (a high-rise hospital, a train yard, an airport, a traffic tunnel, among others) filled with literally thousands of zombies looking to attack you—and even, sometimes, one another. (You want a weird video-game experience? Creep around a corner in the sewers adjacent to the hospital, say, and you might find, to your fascinated horror, a couple of unawares zombies casually *beating each other up*.) These zombies attack singly or in groups or in what the game calls "the horde." Standing in the middle of a darkened city street while a horde of zombies pours up out of a subway station and clamors over and around parked cars to get to you is about as unnerving as video games get. And these are just the rank-and-file zombies. The far more perilous "special infected" is where *Left 4 Dead* begins to glitter.

These special infected come in five nightmare flavors: the Hunter (a hoodied zombie who pounces upon and then tears into his prey, rendering the pouncee helpless until a friend comes along to shoot or push the Hunter off); the Smoker (a coughing, shambolic, elastically tongued zombie who operates much like a sniper, extending his tongue to pluck survivors from the pack); the Boomer (an obese and suppurating slob zombie who is as fragile and explosive as a Pinto but whose vomit and bile attract the dreaded horde, and whose vomit, on top of that, is *blinding*, so that during a well-coordinated attack you cannot see the Hunter tearing to pieces your screaming friend right in front of you); the Tank (as advertised, a steroidally distended zombie as tough as an armored car, but who mercifully appears only a few times a game); and, finally, the Witch (a crying lost-soul zombie who seems the very picture of helplessness, until she is startled by a flashlight or loud noise, upon which she uses her razored manicure to instantly kill the survivor who startled her, and whom you must try to sneak past, and who is as upsetting and inspired a video-game nemesis as any). What is so brilliant about these special infected is the way they tap into distinct types of emotional unease. For the Hunter it is shock and for the Smoker helplessness. For the Boomer it is panic and for the Tank flight. For the Witch it is a strange combination of alarm and paranoia and blame. These emotions, aroused as they are alongside other, living gamers, are part of what makes a game with no traditional narrative to speak of such a dynamically fertile experience to look back on. *Left 4 Dead* creates, within a structure that is formally storyless but highly controlled, a game that feels to those playing it as harrowingly and expertly designed as a first-rate horror film.

Credit here is due to the so-called AI Director that Valve designed specifically for *Left 4 Dead*. It is, most basically, a piece of in-game computation that monitors the gamers, judges their performance, and complicates things as it deems advisable. If things are going really swimmingly for the survivors, why not inflict upon them a Tank? If the survivors are hurting, why not drop in an extra health pack? The AI Director, which could not work in a game with an inflexible narrative structure, also ensures that the survivors are never attacked in the same place by the same number of enemies. The revelatory quality of this innovation cannot be overstated. Gamers often learn how to master a game by memorization, but *Left 4 Dead* is impossible to master in this way. All one can do is hone strategies, which, especially on the highest difficulty level, have a toothpick-house fragility.

You do not get a delivered narrative in *Left 4 Dead*. What you get is a series of found narratives. How do these found narratives in *Left 4 Dead* work, and what gives them their resonance? Well, as it happens, I have a *Left 4 Dead* story and it occurred while playing the game's versus mode, in which two human teams (one survivor, one zombie) have at each other. Playing against human-controlled special infected takes the robotically inflicted havoc of the AI Director and turns it into something far more wonderfully and personally vicious. In versus mode, the object is to reach the safe house with as many living survivors as possible. The more survivors that make it, the more points your team receives. One night, at the end of the first stage of the "Dead Air" campaign, I and three fellow survivors (two of whom were friends, one of whom had just jumped in) had come to realize that we were up against a vilely gifted and absolutely devastating team of *Left 4 Dead* tacticians—the Hannibal, Napoleon, Crazy Horse, and Patton of zombies. They attacked with insurgent coordination and to maximum damage, and it was only our own skill that had managed to hold them off as long as we had. By the time the first-stage safe house came into view, we—four extraordinarily good *Left 4 Dead* veterans—were limping, hobbled, and completely freaked out. Then, another coordinated attack, led by the Boomer puking on us, blinding us, and summoning the horde. While we staggered around, the Smoker took hold of one friend while a Hunter pounced on another. The other remaining survivor and I decided to break for the safe house door. Before getting there my remaining friend was pounced on by yet another Hunter. Although I freed him, I was still mostly blind, and my friend, despite having been released, was under assault by at least a dozen rapacious normal zombies. Deciding that one of us making it was better than none of us making it, I stepped inside the safe house and closed the door. Outside, the friend I had left behind managed to fight his way out of the horde and kill the Smoker and Hunter ripping apart the other survivors, who were now incapacitated, incapable of getting up without help, and quickly bleeding out, which is to say, dying. Unfortunately, the heroic friend was himself incapacitated while doing this. While my three downed friends could shoot their sidearms, they could not rise. They needed me for that. In a minute or so, they would be dead, and from the shelter of the safe house I watched their health bars steadily drain away. Meanwhile, the opposing team had begun to respawn. A lone survivor against even two special infected opponents would stand no chance, as all it would take to end the round would be a Hunter or Smoker incapacitating me. So I stayed put. Better one of us than none of us.

My downed friends failed to see it this way. Over my headphones, they vigorously questioned my courage, my manhood, the ability of my lone female survivor to repopulate the human world on her own, and my understanding of deontological ethics. On the other side of the safe house door, I could hear the Boomer belching, farting, and waiting for me to come out. "You dick!" one of my friends called out. He had just finished bleeding out, a skull appearing beside his onscreen name. My remaining friends were now seconds away from the same fate. I looked within, did not like what I saw, steeled myself, and fired several shotgun rounds through the door, safely killing the Boomer (who it must be said behaved with uncharacteristic carelessness). When I opened the door I saw a Hunter a few feet away, in the corner, waiting to pounce, but I killed him before moving out of the safe room and into the street. The second Hunter was better prepared, but with miraculous good luck I managed to blast him out of the air in mid-pounce. I quickly helped up the first survivor and together we made it out to the final remaining survivor, who was down to his last droplets of virtual existence. While I helped up the final survivor, my friend, covering me, eliminated the lurking Smoker, and with glad cries the three of us made it back into the safe house. At great personal risk, and out of real shame, I had rescued two of my three friends and in the process outfaced against all odds one of the best *Left 4 Dead* teams I had and have ever played against. I realized, then, vividly, that *Left 4 Dead* offered a rare example in which a game's theme (cooperation) was also what was encouraged within the actual flow of gameplay.

The people I saved that night still talk about my heroic action—and, yes, it *was*, it did *feel*, heroic—whenever we play together, and, after the round, two of the opposing team's members requested my online friendship, which with great satisfaction I declined. All the emotions I felt during those few moments—fear, doubt, resolve, and finally courage—were as intensely vivid as any I have felt while reading a novel or watching a film or listening to a piece of music. For what more can one ask? What more could one *want*?

I once raved about *Left 4 Dead* in a video-game emporium within earshot of the manager, a man I had previously heard angrily defend the position that lightsaber wounds are not necessarily cauterized. (His evidence: The tauntaun Han Solo disembowels in *The Empire Strikes Back* does, in fact, bleed.) "*Left 4 Dead*?" he asked me. "You liked it?" I admitted that I did. Very, very much. And him? "I liked it," he said, grudgingly. "I just wished there was more story." A few pimply malingerers, piqued by our exchange, nodded in assent. The overly caloric narrative content of so many games had caused these gentlemen to feel undernourished by the different narrative experience offered by *Left 4 Dead*. They, like the games they presumably loved, had become aesthetically obese. I then realized I was contrasting my aesthetic sensitivity to that of some teenagers about a game that concerns itself with shooting as many zombies as possible. It is moments like this that can make it so dispiritingly difficult to care about video games.

FALLOUT

HEADSHOTS

THE UNBEARABLE
LIGHTNESS OF GAMES

THE GRAMMAR OF FUN

LITTLEBIGPROBLEMS

BRAIDED

MASS EFFECTS

FAR CRIES

GRAND THEFTS

FOUR

E pic Games is a privately owned company and does not disclose its earnings. But on a Monday morning in late April 2008, while standing in Epic's parking lot at Crossroads Corporate Park in Cary, North Carolina, where I was awaiting the arrival of Cliff Bleszinski, Epic's design director, I realized that my surroundings were their own sort of Nasdaq. Ten feet away was a red Hummer H3. Nearby was a Lotus Elise, and next to it a pumpkin-orange Porsche. Many of the cars had personalized plates: PS3CODER, EPICBOY, GRSOFWAR.

Released in late 2006, *Gears of War,* a third-person shooter, was quickly recognized as the first game to provide the sensually overwhelming experience for which the year-old Xbox 360 had been designed. *Gears* won virtually every available industry award and was the 360's best-selling game until Bungie's *Halo 3* came along a year later. Bleszinski, along with everyone else at Epic, was currently "crunching" on *Gears of War 2,* the release date of which was six months away. Its development, long rumored, was not confirmed until the previous February, when, at the Game Developers Conference in San Francisco, Bleszinski made the announcement after bursting through an onstage partition wielding a replica of one of *Gears's* signature weapons—an assault rifle mounted with a chainsaw bayonet.

Bleszinski, who is known to his many fans and occasional detractors as CliffyB, tends to stand out among his colleagues in game design. Heather Chaplin and Aaron Ruby's *Smartbomb* recounts the peacockish outfits and hairstyles he has showcased at industry expos over the years. In 2001 he affected the stylings of a twenty-first-century Tom Wolfe, with white snakeskin shoes and bleached hair. In 2002 he took to leather jackets and an early-Clooney Caesar cut. By 2003, he was wearing long fur-lined coats, his hair skater-punk red. In recent years he let his hair grow shaggy, which gave him the mellow aura of a fourth Bee Gee.

Bleszinski drove into Epic's parking lot in a red Lamborghini Gallardo Spyder, the top down despite an impending rainstorm. His current haircut was short and cowlicked, his bangs twirled up into a tiny moussed horn. He was wearing what in my high school would have been called "exchange-student jeans"—obviously expensive but slightly the wrong color and of a somehow non-American cut. Beneath a tight, fashionably out-of-style black nylon jacket was a T-shirt that read TECHNOLOGY! His sunglasses were of the oversized, county-sheriff variety, and each of his ear-lobes held a small, bright diamond earring. He could have been either a boyish Dolce & Gabbana model or a small-town weed dealer.

Bleszinski suggested that we go to a local diner. He professed an aversion to mornings, and to Monday mornings especially, but seemed dauntingly alert. "This car's like a wake-up call," he said. "By the time I get to work, my heart's pumping and I'm ready to crank." Before we were even out of the parking lot, we were traveling at forty-five miles an hour. At a stoplight, Bleszinski exchanged waves with the driver of an adjacent red Ferrari—another Epic employee. When we hit a hundred miles per hour on a highway entrance ramp, Bleszinski announced, "Never got one ticket!" On the highway, he slowed to seventy-five. "One of my jobs in life," Bleszinski said, cutting over to an exit, "is to make this look a little cooler." By "this," he meant his job. He is adamant that young people interested in gaming should seek to make it their career. After five minutes in Bleszinki's company I was beginning to wonder why I had not attempted to make it *my* career.

Bleszinski's brand of mild outrageousness—the "Cliffycam" on his blog page, which, some years ago, allowed visitors to observe him online while he worked; the photographs of him on his MySpace page alongside the splatter-film director Eli Roth and the porn stars Jenna Jameson and Ron Jeremy—qualifies him as exceptional in an industry that is, as he says, widely assumed to be a preserve inhabited by pale, withdrawn, molelike creatures.

There is some emotional truth to this stereotype. "This industry," Bleszinski told me, "is traditionally filled with incredibly intelligent, talented people who don't necessarily like a lot of attention." In illustration he brought up Bungie's Jason Jones, the primary creative force behind the *Halo* series. "Really, really great guy. But he's shy. He's almost like this Cormac McCarthy–type character. One interview every ten years." When he was young, Bleszinski said, "I wanted to know *who* these people were creating these games. And I was like, 'You know what? If I do a great job making games, maybe people will find it interesting.' "

The CliffyB nickname, he told me, was bestowed on him by some "jock kid," when he was a small, shy teenager, and was meant as a taunt. Bleszinski took the name and fashioned a tougher persona around it, but, after spending a little time with him, I had the sensation of watching someone observing himself. Video games are founded upon such complicated transference. Gamers are allowed, for a time, all manner of ontological assumptions. They can also terminate their assumed personalities whenever they wish, and Bleszinski had lately been asking game-industry journalists if they might not "sit on" the CliffyB moniker "for a while."

Bleszinski was born in Boston, in 1975. His father, whom Bleszinski describes as a "very stressed-out guy," died when he was fifteen. Bleszinski still remembered what game he was playing when he learned of his father's death: the Nintendo game *Blaster Master.* He never played it again.

In 1991 Bleszinski's mother bought him a computer. "I'm not that technical of a guy," Bleszinski told me. "I started off programming my own games and doing the art for them, but I was a crappy coder and a crappier artist. But I didn't want to let that stop me. I was going to kick and scream and claw my way into the business any way possible." He thinks that it was perhaps only the death of his father that allowed him to do this. "If he hadn't passed, he probably would have made me go to Northeastern and become an engineer."

In 1992, a year after Bleszinski had taught himself the rudiments of computer programming, he sent a game submission to Tim Sweeney, the twenty-two-year-old CEO of Epic MegaGames, who had recently dropped out of a mechanical-engineering program at the University of Maryland. (Epic's original name, Sweeney told me, was "a big scam" to make it look legitimate. "When you're this one single person in your parents' garage trying to start a company, you want to look like you're really big.") Sweeney fondly recalled Bleszinski's submission, a PC game called *Dare to Dream,* in which a boy gets trapped, appropriately enough, inside his own dreamworld. "At that time," Sweeney said, "we'd gotten sixty or seventy game submissions from different people and we went ahead with five of them, and Cliff was one of the best." Once his game was in development, Bleszinski became more involved with the company. Sweeney said, "We'd have four or five different projects in development with different teams, and every time one started to reach completion we'd send it off to Cliff, and he'd write off his big list of what's good about this game and what needs to be fixed. Cliff's lists kept growing bigger and bigger." Bleszinski was only seventeen. When I asked Sweeney if he had had any reservations about entrusting his company's fate to a teenager whose driver's-license lamination was still warm, Sweeney said, "That's what the industry was like."

Upon its release, *Dare to Dream,* in Bleszinski's words, "bombed." His second game for Epic, released a year later, featured what he calls a "Rambo rabbit" named Jazz, who carried a large gun and hunted frenetically for intergalactic treasure. *Jazz Jackrabbit,* which imported to a PC platform a type of gameplay previously exclusive to Sega and Nintendo consoles, made Bleszinski's reputation. Epic had begun as a company churning out somewhat réchauffé fare—pinball simulators, clone-ish retreads, but also a popular shooter known as *Unreal Tournament,* the updated and augmented engine of which has since been adopted by hundreds of games. It was Bleszinski, along with the Unreal Engine, that helped to fill Epic's parking lot with sports cars.

At the diner I asked Bleszinski which games had most influenced him. *Super Mario Bros.,* he said, "is where it really kicked into high gear." Indeed, the first issue of *Nintendo Power,* published in 1988, listed the high scores of a handful of *Super Mario* devotees, the thirteen-year-old Bleszinski's among them. "There's something about the whole hidden element to *Mario,*" Bleszinski said, "where you jump and hit your head on a block and just out of nowhere secret things would appear. They made you feel like a kid in the woods finding god knows what."

Released in 1985, *Super Mario Bros.* was a game of summer-vacation-consuming scope and unprecedented inventiveness. It was among the first video games to suggest that it might contain a world. It was also hallucinogenically strange. Why did mushrooms make Mario grow larger? Why did flowers give Mario the ability to spit fire? Why did bashing Mario's head against bricks sometimes produce coins? And why was Mario's enemy, Bowser, a saurian, spiky-shelled turtle?

In film and literature, such surrealistic fantasy typically occurs at the outer edge of experimentalism, but early video games depended on symbols for the simple reason that the technological limitations of the time made realism impossible. Mario, for instance, wore a porkpie hat not for aesthetic reasons but because hair was too difficult to render. Bleszinski retains affection for many older games, but he says, "If you go back and play the majority of old games, they really aren't very good." He suspects that what made them seem so good at the time was the imaginative involvement of players: "You wanted to believe, you wanted to fill in the gaps."

A game like *Gears of War* differs so profoundly from *Super Mario Bros.* that the two appear to share as many commonalities as a trilobite does with a Great Dane. *Super Mario* requires an ability to recognize patterns, considerable hand–eye coordination, and quick reflexes. *Gears* requires the ability to think tactically and make subtle judgments based on scant information, a constant awareness of multiple variables (ammunition stores, enemy weaknesses) as they change throughout the game, and the spatial sensitivity to control one's movement through a space in which the "right" direction is not always apparent. Anyone who plays modern games such as *Gears* does not so much learn the rules as develop a kind of intuition for how the game operates. Often, there is no single way to accomplish a given task; improvisation is rewarded. Older games, like *Super Mario,* punish improvisation: You live or die according to their algebra alone.

Gears is largely the story of a soldier, Marcus Fenix, who, as the game begins, has been imprisoned for abandoning his comrades (who are known as COGs, or Gears) in order to save his father, who dies before Fenix can reach him. He is released because a fourteen-year war with a tunnel-using alien army known as the Locusts has depleted the human army's ranks. This much we glean in the first two minutes of the game; the next ten hours or so are an ingeniously paced march through frequent and elaborately staged firefights with Locusts, Wretches, Dark Wretches, Corpsers, Boomers, three blind and terrifying Berserkers, and the vile General Raam. Along the way, players can treat themselves to the singular experience of using the chainsaw bayonets on their Lancer assault rifles to cut their enemies in half, during which the in-game camera is gleefully splashed with blood. (*Gears* is one of the most violent games ever made, but Bleszinski maintains that it contains "very much a laughable kind of violence," like "watching a melon explode in a Gallagher show.")

The story line and the narrative dilemmas of *Gears* are not very sophisticated. What is sophisticated about *Gears* is its mood. The world in which the action takes place is a kind of destroyed utopia; its architecture, weapons, and characters are chunky and oversized but, somehow, never absurd. Most video-game worlds, however well conceived, are essenceless. *Gears* feels dirty and inhabited, and everything from the mechanics of its gameplay to its elliptical backstory has been forcefully conceived, giving it an experiential depth rare in the genre.

Much of the resonance of *Gears* can be directly attributed to the character of Fenix. Video-game characters tend to be emptily iconic. Pac-Man, for instance, is some sort of notionally symbolic being. Mario (who was originally known only as Jumpman) is merely the vessel of the player's desired task. *Tomb Raider*'s Lara Croft is either the embodiment of or a satire on female objectification. *Halo*'s Master Chief is notable mainly for his golden-visored unknowableness. Fenix, encased within armor so thick that his arms and legs resemble hydrants, his head covered by a black bandanna, and his eyes as tiny as BBs, is different. He shows constant caution and, occasionally, fear. Although he can dive gracefully, his normal gait has the lumbering heaviness of an abandoned herd animal. His face is badly scarred, and his voice is a three-packs-a-day growl less angry than exhausted. Unlike the protagonists of many shooters, the Fenix of the first *Gears* rarely seems particularly eager to kill anything. The advertising campaign for *Gears of War* was centered on a strangely affecting sixty-second spot in which Fenix twice flees from enemies, only to be cornered by a Corpser, a monstrous arachnid creature, on which he opens fire. But it was the sound track—Gary Jules's spare, mournful cover of the 1982 Tears for Fears song "Mad World"—that gave the spot its harsh–tender dissonance. This helped signal that Fenix was something few video-game characters had yet managed to be: disappointedly adult.

When I asked about the melancholy at the core of *Gears,* Bleszinski said, "I was never geeky enough for the geeks and I was never cool enough for the cool people. I've always been in that weird purgatory." That slight feeling of identity crisis apparently persisted. A few years ago, Bleszinski divorced his high school sweetheart. "I woke up one day, and had the two Labs, and the house in the suburbs, and I'm like, What the hell am I doing here?"

Bleszinski admitted that much of *Gears* is, in its way, autobiographical. Its look and its aesthetic, for example, were influenced by his first trip to London, taken when he was in his late twenties.

There he climbed to the top of St. Paul's Cathedral and, with "a shitty little camera," snapped a picture of a yolky sun setting over the Thames, the sky streaked with nursery blues and pinks. Bleszinski's London photograph is one of the reasons that much of *Gears* takes place in twilight—a lighting condition prized by cinematographers but comparatively neglected in video games. Bleszinski asked his artists to create a "sci-fi" hybrid of London and Washington, DC, but advised them to keep the futuristic well balanced with the historical. The big flaw in most depictions of the future, he says, "is that they always forget to leave in the past. Everyone always assumes that the entire world would just explode and be rebuilt in this kind of super-futuristic style. I still see old cars from the '30s and '40s around, right next to things that look like they're from the year 2000. It's that mix that makes things interesting."

Gears also contains what Bleszinski calls a "going home" narrative: "There's a sublevel to *Gears* that so many people missed out on because it's such a big testosterone-filled chainsaw-fest. Marcus Fenix goes back to his childhood home in the game. I dream about my house in Boston, basically every other night. It was up on a hill." In *Gears of War* the fatherless Fenix's manse is on a hill, too, and getting to its front door involves some of the most harried and ridiculously frantic fighting in the game. When I told Bleszinski that Fenix's homecoming was one of my favorite levels in *Gears,* he asked if I knew where its title, "Imaginary Place," had come from. I thought for a moment, attuned to the possibility of an altogether unexpected window into his imagination. Was it from Auden? No. It was a reference to a line from Zach Braff's film *Garden State,* in which *family* is defined as "a group of people who miss the same imaginary place." When you start to peel back the layers of the *Gears* world, Bleszinski told me, "there's a lot of sadness there." Indeed, the centaur tanks the COGs ride into battle in

Gears of War 2 greatly resemble the tanks from *Blaster Master*—the game Bleszinski was playing when he learned of his father's death—which Bleszinski claimed not to have realized until someone else pointed it out to him.

In addition to the prevailing mood of wistful savagery, the singularity of *Gears of War* resides in the "feel" of its game mechanics. (The procedures and rules of a game are what are meant, broadly speaking, by the term *game mechanics*. As the game designer Jesse Schell writes, "If you compare games to more linear entertainment experiences [books, movies, etc.], you will note that while linear experiences involve technology, story, and aesthetics, they do not involve mechanics, for it is mechanics that make a game a game.") The importance of game-mechanic feel is something that Bleszinski has made his special focus and passion. "I'm looking for a fun core-loop of what you're doing for thirty seconds over and over again," he told me. "I want it to grab me quick and fast. I want it to have an interesting game mechanic, but I also want it to be a fascinating universe that I want to spend time in, because you're spending often dozens of hours in this universe." The best example of "an interesting game mechanic" is *Gears*'s take on the hoary conceit of reloading one's weapon: A well-timed reload briefly rewards the gamer with enhanced damage infliction. Rod Fergusson, an Epic senior producer who, with Bleszinski, oversees the continuity of the *Gears* universe, told me that Bleszinski is "a designer by feel" who conceives of games in "big-picture" terms "yet tweaks the smallest things." He said, "If you look at people who tried to copy *Gears*'s mechanics, they don't have that guy doing that hands-on, touchy-feely way of designing. They kind of get the broad strokes, but they don't get the little detaily things."

Evidence of this can be detected in *Gears*'s famous "cover" system, which demands that the player move with chesslike care and efficiency around the battle space, using walls, doorways, barricades, and the scorched husks of vehicles as cover. Bleszinski told me that a paintball match had impressed upon him the ludicrousness of how most shooters operated, with players running around in the open, strafing their enemies and jumping to avoid being shot. It occurred to him that a shooter based on the idea of taking cover would be a more realistic and primal experience. Namco's 2003 shooter *kill.switch* was the first game to attempt a cover system, but Bleszinski, an adroit borrower, streamlined and improved the concept. In *Gears,* an exposed player is actively punished by the game, as the damage inflicted by enemy bullets and explosives spikes nastily whenever cover is abandoned. Bleszinski became fixated on making sure that, when cover is taken, the right amount of dust is kicked up against the controlled character's back and that the character's grunt has just the right timbre and volume. In this way, the reward for seeking cover becomes subliminally sensory.

The most frequently imitated aspect of *Gears* is a feature known as the "roadie run," so named for the crouch into which a player's hustling character lowers himself, and which Bleszinski thought resembled a rock-show roadie's attempt to move discreetly across the stage. While in roadie-run mode, the in-game camera goes jitteringly handheld and fish-eyed and sinks into a miasma of dust. It is difficult to see exactly where one is going, and the overall effect is that of intense panic. Bleszinski calls it "the Falluja followcam," and likens it to the viewpoint of an embedded journalist. Yet in roadie run the player is traveling only one and a half times faster than normal. The feature is both a brilliant distortion of perspective and a cunning approximation of the confusion of combat.

If one were to commission a very bright and unusually tasteful adolescent to design his ideal workplace, Epic's headquarters would probably be the result. Its many blacks, grays, and corrugated-metal surfaces might best be labeled Bachelor Futurist, even though, by now, most of Epic's male employees wear wedding rings. The office furnishings come in three styles: Neo–Living Room (easy-rocking, lever-activated recliners), Casual Satanist (black leather couches), and Romper Room Gothic (beanbag chairs). The aroma of lingering adolescence carries over into the mementos, knickknacks, and emblems that Epic's employees use to decorate their offices. Tim Sweeney, with the earnestness of a teenage boy, has a prominent Ferrari flag hanging in his office—though Sweeney, unlike most teenage boys, actually owns a Ferrari. Chris Perna, the art director of *Gears 2,* displays on one of the shelves in his office a foot-high silver-cast Darth Vader Pez dispenser. Bleszinski's office resembles a toy-store yard sale. These are boyish affectations, certainly, but boyishness is the realm in which these men seek inspiration, not a code by which they live.

A Microsoft employee who works closely with Epic described the company as having a "band dynamic." Staff turnover is low, and many of Epic's most senior employees have been friends for more than a decade. This does not seem a very long time until one sits in on an Epic meeting and realizes that anyone over the age of thirty-five achieves the temporal stature of Methuselah. Epic's recent growth is regarded with wary gratitude by many of its employees, though some miss the old days, when, as Sweeney put it, "we were just a bunch of kids who had some cool ideas and were doing neat things."

When surrounded by his colleagues and discussing the gravities of gameplay, Bleszinski discarded his self-consciously laidback manner, and the precision of his gaming mind quickly became apparent. A colleague told me that Bleszinski's "huge strength is his basic ability to just get it—pick something up and give you a one-minute usability report."

It is unusual for any game company to allow an outsider access to its meetings, for fear of a game's features being prematurely disclosed. While discussing *Gears 2*'s new "crowd" system, which allows an unprecedented number of individually functioning enemies to flock across the battle space, Bleszinski mentioned how excited he was to open fire upon them with *Gears 2*'s mortar. Within minutes, I was pulled aside by a Microsoft representative and informed that the mortar's existence would not be confirmed until later in the summer and could I please refrain from mentioning it to anyone. (That I had once been allowed to sit in on classified intel meetings while embedded with the Marine Corps in Iraq did not, at Epic, carry much weight.) The gaming media is largely made up of obsessive enthusiasts, and the carefully planned release of information tantalizes them with the promise of insider knowledge. The game industry is more or less leakproof and possesses a strange kind of innocence: It guards its secrets as guilelessly as a boy might hide from his mother—but not from his brother and sister—the extraterrestrial living in his bedroom.

I was warned, before attending a play-through of one of *Gears 2*'s unfinished levels, that it was still "janky." Although the level was fully voice-acted, it was only partially scored, and many sound effects had yet to be added. Some of the virtual lighting was not yet functional, and the onscreen Fenix flickered. From time to time, the game crashed altogether. Nonetheless, we watched Dave Nash, the level's lead designer, guide Fenix through what looked to be an enthusiastically mortared office building. Ahead, in the shadows, numerous monsters scrambled—the goose-bumpy *Gears* warning that a violent engagement was somewhere ahead. By the third iteration of this, Bleszinski was shaking his head. "Enough monster foreplay." When Fenix and a comrade had to walk a bomb to a door in need of obliteration, Bleszinski said that they should be moving "10 to 15 percent faster." When the play-through was complete, he said that too much of the level involved going into one room to hit a switch that activated a door in another. The otherwise superlative *Grand Theft Auto* series is particularly afflicted with this mouse-coaxed-through-a-maze problem, and Bleszinski's only complaint about the *GTA* games, which he admires, is that they can sometimes "feel like work." This emphasis on making players aware of why they were choosing certain paths without being reminded that their choice was essentially illusory made me think of the "vivid and continuous dream" that John Gardner once spoke of concerning fiction.

In a discussion of another level of the game, known as "Hospital," Bleszinski, sitting by himself on a long couch, wondered if, in the online multiplayer version of the level, a hospital sprinkler system could be used to extinguish the pilot light of an opponent's flamethrower. "Wouldn't that be cool?" he asked. A similar moment resulted in one of *Gears 1*'s most notorious features. One day, late in the game's production, Bleszinski expressed his frustration at having to finish off downed enemies by shooting them. "You know," he said, "I just want to stomp on his head!" From this came the game's gratuitously fatal coup de grâce: the "curb stomp," whereby a player crushes his opponent's skull beneath the anvil of an enormous metal boot. Rod Fergusson later told me, "I wouldn't be surprised if, in the next two weeks, there's a hallway with a sprinkler system that puts out the pilot light of the flamethrower." (And there would be.)

Discussion at Epic is collegial and to the point; modern game design is too complex and collaborative for any individual to feel proprietary about his or her own ideas. At one meeting I attended, a disagreement about weaponry was swiftly resolved. "There's no direct counter to the flamethrower," Ray Davis, the game's lead programmer, pointed out, with exasperation. Lead gameplay designer Lee Perry, who had obviously heard this before, sighed. "I don't know. I think it's a superweapon," he said. Then someone else observed that the boomshot, another devastatingly fatal weapon, had no direct counter, either, and Davis recognized with a grin that his argument had been destroyed. Bleszinski took the opportunity to raise a singular annoyance of the boomshot, familiar to anyone with experience of the multiplayer version of *Gears:* the impossibility of knowing whether someone you are charging toward happens to be carrying a boomshot. "The boomshot needs something to *warn* you your opponent's got it," Bleszinski said. He suggested adding small glowing lights around its four barrels, which everyone agreed was a fine solution. Davis, who worked most directly with the programmers and was therefore most familiar with what remained janky, brought up the "inconsistent, unfun lethality" of frag grenades. This segued into *Gears 2's* inclusion of ink grenades, which create a highly damaging toxic cloud—the proper gameplay use of which no one, so far, had been able to decide.

Bleszinski and the other Epic designers came to this form as children. Growing up playing games, they absorbed the governing logic of the medium, but no institutions existed for them to transform what they learned into a methodology. Gradually, though, they turned a hobby into a creative profession that is now as complex as any other. I realized, watching them, that part of what they had done was help to establish the principles of one grammar of fun.

Before leaving Epic, I was invited to take part in the daily playtest, which occupies an hour or two of every afternoon. That day the team was testing the multiplayer modes of *Gears 2.* One of the most common criticisms of video games is that they can wrap those who play in enforced and occasionally deranging solitude, but to take part in a multiplayer game is to give a game new life every time one plays, because one is matched against human players, whose ingenuity and deviousness no computer can hope to equal, and because one can exchange with one's fellow players advice, congratulations, and taunts (mostly taunts). A dozen Epic employees gathered in the test lab and signed in to the individual consoles that lined the walls. The battle would pit one side of the room against the other. I was assigned my slot and selected my avatar—a Drone Locust.

Testing is among the more consuming aspects of modern game design. I was told that *Gears 2* would be subject to roughly forty thousand hours of testing before its release. This is an impressive number—until one realizes that the first forty thousand people who buy and play *Gears 2* will, in one hour, equal that amount of testing. *Gears 1* was released with a few bugs—instances in which the game behaves in a way that was not intended—a fact that many at Epic remain bitterly embarrassed about, even though it is nearly impossible to completely eliminate bugs from any game. Testing of multiplayer modes involves something more sociological than purely technical assessment: learning what tendencies a given environment fosters. If there are places to hide, they must always have a fatal weakness. If a particularly powerful weapon is hidden somewhere, it must be difficult and risky to reach. If large numbers of players are being killed at a certain location, the game designers must ask themselves why, and decide whether to correct for this. The aim is to eliminate all ineradicable advantages, but this goal is seldom attained. Two weeks after the release of the first *Gears*, Bleszinski told me, "I'd go online and get completely destroyed by everybody."

In our first multiplayer game, which was called "Guardian," one had to kill the opposing team's leader and all those who protect him. By the end, a teammate and I happened to be the only survivors (I achieved this status largely by hiding), and we encountered Bleszinski crouching behind a stack of sandbags. We decided to charge. Bleszinski popped out from cover and, with a shotgun, expertly exploded the head of my teammate before beating me to death as I rounded the edge of his hiding place. Bleszinski ended the game with twenty-one kills; I had three.

The next contest, then known as "Meatflag" but since renamed, amounted to a game of capture the flag—though the flag was, bracingly, a human being. In this match, I was simultaneously chainsawed to death by three people, a spectacle that everyone in the room claimed never to have seen before, in all their hours of play. Our final game was called "Wingman," which is played in pairs. Bleszinski and I buddied up, and I shouted across the room to him for some general guidance. "Basically," he said, "kill anyone who doesn't look like you. Our foreign policy."

I fared miserably again, and pride compelled me to point out that I had finished *Gears 1* on its most challenging difficulty level. No one was listening, and Bleszinski stood up. "Now's the fun part," he said. "Figure out what's a bug and what's not a bug." He conferred in a huddle with the other designers about what to enter into the defect-tracking database. "You tell people what you do for a living," Bleszinski said later, "and they're like, 'Oh, you play video games for a living.' No, I play a game that's not as fun as it should be, that's broken, until it's no longer broken. Then I give it to other people to have fun with."

FALLOUT

HEADSHOTS

THE UNBEARABLE
LIGHTNESS OF GAMES

THE GRAMMAR OF FUN

LITTLEBIGPROBLEMS

BRAIDED

MASS EFFECTS

FAR CRIES

GRAND THEFTS

FIVE

To learn what the video-game industry at large thought of itself and where it believed it was going, I went to Las Vegas, a city to which I had moved two years earlier for a ten-month writing fellowship. I had not expected to enjoy my time in Vegas but, to my surprise, I did. I liked the corporate diligence with which upper-tier prostitutes worked the casino bars and the recklessness with which the Bellagio's fountains blasted the city's most precious resource into the air a dozen times a day, often to the chorus of "Proud to Be an American." Some days I sat on my veranda and watched the jets float in steady and low over the city's east side, bringing in the ice-encased sushi and the Muscovite millionaires and the husky midwesterners and the collapsed-star celebrities booked for a week at the Mirage. I even liked the sense I had while living in Las Vegas that what separated me from a variety of apocalyptic ruins was nothing more than a few unwise decisions.

Las Vegas itself is as ultimately doomed as a colony of sea monkeys. One vexation is water, of which it is rapidly running out. Another is money, of which it needs around-the-clock transfusions. The city's murder-suicide pact with its environment and itself is in-built, congenital. Constructed too shoddily, governed too erratically, enjoyed and abused by too many, Las Vegas was the world's whore, and whores do not change. Whores collapse.

Collapsing was what Las Vegas in the winter of 2009 seemed to be doing. The first signs were small. From my rental car I noticed that a favorite restaurant had a sign that read RECESSION LUNCH SPECIAL. Laundromats, meanwhile, promised FREE SOAP. More ominously, one of Vegas's biggest grocery store chains had gone out of business, resulting in several massive, boarded-up complexes in the middle of stadium-sized parking lots, as indelible as the funerary temples of a fallen civilization. Entire office parks had been abandoned down to their electrical outlets. Hand-lettered SAVE YOUR HOUSE signs marked every other intersection, while other signs, just below them, offered FORECLOSURE TOURS. At one stoplight a GARAGE SALE BEHIND YOU notice turned me around. I found a nervous middle-aged white woman selling her wedding dress ($100) and a small pile of individual bookcase shelves ($1). She smiled hopelessly as I considered her wall brackets ($.15) and cracked flowerpots ($.10), all set out on an old card table ($5).

This was not the Vegas I remembered, but then most of my time there was spent playing video games. A game I played only because I lived in Vegas was Ubisoft's shooter *Rainbow Six Vegas 2,* one of many iterations of a series licensed out in the name of the old scribbling warhorse Tom Clancy. *Rainbow Six Vegas 2* is mostly forgettable, though it is fun to fight your way through the Las Vegas Convention Center and take cover behind a bank of Las Vegas Hilton slot machines. It was also fascinating to see the latest drops of conceit wrung from *Rainbow Six*'s stirringly improbable vision of Mexican terrorists operating with citywide impunity upon the American mainland. The game's story is set in 2010. While no one will be getting flash-banged in the lobby of Mandalay Bay anytime soon, driving around 2009 Las Vegas made the game's casino gunfights and the taking of UNLV seem slightly less unimaginable.

Out at Vegas's distant Red Rock hotel and casino, the Academy of Interactive Arts & Sciences was throwing its annual summit, known as DICE (Design Innovate Communicate Entertain), which gathers together—for the purpose of panels, networking, an awards show, and general self-celebration—the most powerful people in the video-game industry. With the Dow torpedoed, layoffs occurring in numbers that recall mass-starvation casualties, and newspapers and magazines closing by the hour (including the game-industry stalwart *Electronic Gaming Monthly*), DICE held out the reassurance of mingling with the dukes and (rather more infrequently) duchesses of a relatively stable kingdom—though it, too, had been bloodied. Electronic Arts, the biggest video-game publisher in the world, lost something like three-quarters of a billion dollars in 2008. Midway, creators of the sanguinary classic fighting game *Mortal Kombat* and one of the few surviving game developers that began in the antiquity of the Arcade Age, had been recently sold for quite a bit less than a three-bedroom home on Lake Superior. One of the still-unreleased games Midway (with Surreal Software) had spent tens of millions of dollars in recent years developing is called *This Is Vegas,* an open-world game in the *Grand Theft Auto* mode that, according to some promotional material, pits the player against "a powerful businessman" who wants to turn Vegas "into a family-friendly tourist trap." The player, in turn, must fight, race, gamble, and party his or her "way to the top." In today's Las Vegas, the only thing one could hope to party his way to the top of is the unemployment line, and the game's specter of a "family-friendly" city seemed suddenly, even cruelly, obtuse.

Upon check-in every DICE attendee received a cache of swag that included a resplendent laptop carrying case, an IGN.com water bottle (instructively labeled HANG OVER RELIEF), a handsome reading light–*cum*–bookmark, the latest issue of the industry trade magazine *Develop* (President Obama somehow made its cover, too), and a paperback copy of a self-help business book titled *Super Crunchers*. Shortly after receiving my gift bag, I ran into a young DICE staffer named Al, who responded to my joke about the Obama cover ("Yes Wii Can") by reminding me of the Wii that currently occupied the White House rec room. "By 2020," Al told me, "there is a very good chance that the president will be someone who played *Super Mario Bros.* on the NES." I had to admit that this was pretty generationally stirring. The question, I said, was whether the 2020 president-elect would *still* be playing games. Maybe he would. The "spectacle" of games, Al told me, was on its way out. Increasingly important, he said, was "message."

Many have wondered why a turn toward maturity has taken the video game so long. But has it? Visual mediums almost always begin in exuberant, often violent spectacle. A glance at some of the first, most popular film titles suggests how willing film's original audiences were to delight in the containment of anarchy: *The Great Train Robbery, The Escaped Lunatic, Automobile Thieves*. Needless to say, a film made in 1905 was nothing like a film made twenty years later. Vulturously still cameras had given way to editing, and actors, who at the dawn of film were not considered proper actors at all, had developed an entirely new, medium-appropriate method of feigned existence. Above all, films made in the 1920s were responding to other films—their blanknesses and stillnessess and hesitations. While films became more *formally* interesting, video games became more *viscerally* interesting. They gave you what they gave you before, only more of it, bigger and better and more prettily rendered. The generation of game designers currently at work is the first to have a comprehensive growth chart of the already accomplished. No longer content with putting better muscles on digital skeletons, game designers have a new imperative—to make gamers *feel* something beyond excitement.

One designer told me that the idea of designing a game with any lasting emotional power was unimaginable to him only a decade ago: "We didn't have the ability to render characters, we didn't know how to direct the voice acting—all these things that Hollywood does on a regular basis—because we were too busy figuring out how to make a rocket launcher." After decades of shooting sprees, the video game has shaved, combed its hair, and made itself as culturally presentable as possible. The sorts of fundamental questions posed by Aristotle (what is dramatic motivation? what is character? what does story *mean*?) may have come to the video game as a kind of reverse novelty, but at least they had finally come.

At DICE, one did not look at a room inhabited by video-game luminaries and think, *Artists*. One did not even necessarily think, *Creative types*. They looked nothing like gathered musicians or writers or filmmakers, who, having freshly carved from their tender hides another album or book or movie, move woundedly about the room. But what does a "game developer" even look like? I had no idea. The "game moguls" I believed I could recognize, but only because moguls tend to resemble other moguls: human stallions of groomed, striding calm. A number of DICE's hungrier attendees wore plush velvet dinner jackets over *Warcraft* T-shirts, looking like youthful businessmen employed by some disreputably edgy company. There was a lot of vaguely embarrassing sartorial showboating going on, but it was hard to begrudge anyone that. Most of these people sit in cubicle hives for months, if not years, staring at their computer screens, their medium's governing language—with its "engines" and "builds" and "patches"—more akin to the terminology of auto manufacture than a product with any flashy cultural cachet. (In actual fact, the auto and game industries have quite a bit in common. Both were the unintended result of technological breakthroughs, both made a product with unforeseen military applications, and both have been viewed as a public safety hazard.)

There was another kind of DICE attendee, however, and he was older, grayer, and ponytailed—a living reminder of the video game's homely origins, a man made phantom by decades of cultural indifference. An industry launched by burrito-fueled grad school dropouts with wallets of maxed-out credit cards now had groupies and hemispherical influence and commanded at least fiduciary respect. Was this man relieved his medium's day had come or sad that it had come now, so distant from the blossom of his youth? It was surely a bitter pill: The thing to which he had dedicated his life was, at long last, cool, though he himself was not, and never would be.

Like any complicated thing, however, video games are "cool" only in sum. Again and again at DICE, I struck up a conversation with someone, learned what game they had done, told them I loved that game, asked what they had worked on, and been told something along the lines of, "I did the smoke for *Call of Duty: World at War*." Statements such as this tended to freeze my conversational motor about as definitively as, "I was a concentration camp guard."

Make no mistake: Individuals do not make games; *guilds* make games. *Technology* literally means "knowledge of a skill," and a forbidding number of them are required in modern game design. An average game today is likely to have as much writing as it does sculpture, as much probability analysis as it does resource management, as much architecture as it does music, as much physics as it does cinematography. The more technical aspects of game design are frequently done by smaller, specialist companies: I shook hands with the CEO of the company who did the lighting in *Mass Effect* and chatted with another man responsible for the facial animation in *Grand Theft Auto IV.*

"Games have gotten a lot more glamorous in the last twenty years," one elder statesman told me ruefully. Older industry expos, he said, usually involved four hundred men, all of whom took turns unsuccessfully propositioning the one woman. At DICE there were quite a few women, all of whom, *mirabile dictu,* appeared fully engaged with rampant game talk. At the bar I heard the following: Man: "It's not your typical World War II game. It's not storming the beaches." Woman: "Is it a stealth game, then?" Man: "More of a run-and-gun game." Woman: "There's stealth elements?" The industry's woes often came up. When one man mentioned to another a mutual friend who had recently lost his job, his compeer looked down into his Pinot Noir. "Lot of movement this year," he said grimly. Fallen comrades, imploded studios, and gobbled developers were invoked with a kind of there-but-for-the-power-up-of-God-go-we sadness.

Many had harsh words for the games press. "They don't review for anyone but themselves," one man told me. "Game reviewers have a huge responsibility, and they abuse it." This man designed what are called "casual games," which are typically released for handheld systems such as the Nintendo DS or PSP. In many cases developer royalties are attached to their reviewer-dependent Metacritic scores, and because game journalists can be generally relied upon to overpraise the industry's attention-hoarding AAA titles (shooters, RPGs, fighting games, and everything else aimed at the eighteen-to-thirty-four male demographic—a lot of which games I myself admire), the anger from developers who worked on smaller games was understandable. Another man introduced himself to tell me that, in four months, his company would release its first game on Xbox Live Arcade, the online service that allows Xbox 360 owners access to a growing library of digitally downloaded titles. This, he argued, is the best and most sustainable model for the industry: small games, developed by a small group of people, that have a lot of replay value, and, above all, are *fun*. According to him, pouring tens of millions into developing AAA retail titles is part of the reason why the EAs of the world are bleeding profits. The concentration on hideously expensive titles, he said, was "wrong for the industry." (For one brief moment I thought I had wandered into a book publishing party.)

Eventually I found myself beside Nick Ahrens, a choirboy-faced editor for *Game Informer*, which is one of the sharpest and most cogent magazines covering the industry. "These guys," Ahrens said, motioning around the room, "are using their childhoods to create a *business*." The strip-mining of childhood had taken video games surprisingly far, but childhood, like every natural resource, is exhaustible.

DICE's first panel addressed the tricksy matter of "Believable Characters in Games." As someone whose palm frequently seeks his forehead whenever video-game characters have conversations longer than eight seconds, I eagerly took my seat in the Red Rock's Pavilion Ballroom long before the room had reached even 10 percent occupancy. The night before there had been a poker tournament, after which a good number of DICE attendees had carousingly traversed Vegas's great indoors. Two of my three morning conversations had been like standing at the mouth of a cave filled with a three-hundred-year-old stash of whiskey, boar meat, and cigarettes.

"Believable characters" was an admirable goal for this industry to discuss publicly. It was also problematical. For one thing, the topic presupposed that "believability" was quantifiable. I wondered what, in the mind of the average game designer, believability actually amounted to. Oskar Schindler? Chewbacca? Bugs Bunny? Because video-game characters are still largely incapable of actorly nuance, they frequently resemble cartoon characters. Both are designed, animated, and artisanal—the exact sum of their many parts. But games, while often cartoonish, are not cartoons. In a cartoon, realism is not the problem because it is not the goal. In a game, frequently, the opposite is true. In a cartoon, a character is brought to life independent of the viewer. The viewer may judge it, but he or she cannot affect it. In a game, a character is more golemlike, brought to life first with the incantation of code and then by the gamer him- or herself. Unlike a cartoon character, a video-game character does not inhabit closed space; a video-game character inhabits *open situations*. For the situations to remain compelling, some strain of realism—however stylized, however qualified—must be in evidence. The modern video game has generally elected to submit such evidence in the form of graphical photorealism, which is a method rather than a guarantee. By mistaking realism for believability, video games have given us an interesting paradox: the so-called Uncanny Valley Problem, wherein the more lifelike nonliving things appear to be, the more cognitively unsettling they become.

The panel opened with a short presentation by Greg Short, the co-founder of Electronic Entertainment Design and Research. What EEDAR does is track industry trends, and according to

Short he and his team have spent the last three years researching video games. (At this, a man sitting next to me turned to his colleague and muttered, "This can't be a good thing.") Short's researchers identified fifteen thousand attributes for around eight thousand different video-game titles, a task that made the lot of Tantalus sound comparatively paradisaical. Short's first Power-Point slide listed the lead personas, as delineated by species. "The majority of video games," Short said soberly, "deal with human lead characters." (Other popular leads included "robot," "mythical creature," and "animal.") In addition, the vast majority of leading characters are between the ages of eighteen and thirty-four. Not a single game EEDAR researched provided an elderly lead character, with the exception of those games that allowed variable age as part of in-game character customization, which in any event accounted for 12 percent of researched games. Short went on to explain the meaning of all this, but his point was made: (a) People like playing as people, and (b) They like playing as people that almost precisely resemble themselves. I was reminded of Anthony Burgess's joke about his ideal reader as "a lapsed Catholic and failed musician, short-sighted, color-blind, auditorily biased, who has read the books that I have read." Burgess was kidding. Mr. Short was not, and his presentation left something ozonically scorched in the air. I thought of all the games I had played in which I had run some twenty-something masculine nonentity through his paces. Apparently I had even more such experiences to look forward to, all thanks to EEDAR's findings. Never in my life had I felt more depressed about the democracy of garbage that games were at their worst.

The panel moderator, Chris Kohler, from *Wired* magazine, introduced himself next. His goal was to walk the audience through the evolution of the video-game character, from the australopithecine attempts (*Pong's* roving rectangle, *Tank's* tank) to the always-interesting Pac-Man, who, in Kohler's words, was "an abstraction between a human and symbol." Pac-Man, Kohler explained, "had a life. He had a wife. He had children." Pac-Man's titular Namco game also boasted some of the medium's first cut scenes, which by the time of the game's sequel, *Ms. Pac-Man*, had become more elaborate by inches, showing, among other things, how Mr. and Ms. Pac-Man met. "It was not a narrative," Kohler pointed out, "but it was giving life to these characters." Then came Nintendo's *Donkey Kong*. While there was no character development to speak of in *Donkey Kong* ("It's not Mario's journey of personal discovery"), it became a prototype of the modern video-game narrative. In short, someone wanted something, he would go through a lot to get it, and his attempts would take place within chapters or levels. By taking that conceit and bottlenecking it with the complications of "story," the modern video-game narrative was born.

How exactly this happened, in Kohler's admitted simplification, concerns the split between Japanese and American gaming in the 1980s. American gaming went to the personal computer, while Japanese gaming retreated largely to the console. Suddenly there were all sorts of games: platformers, flight simulators, text-based adventures, role-playing games. The last two were supreme early examples of games that, as Kohler put it, have "human drama in which a character goes through experiences and comes out different in the end." The Japanese made story a focus in their growingly elaborate RPGs by expanding the length and moment of the in-game cut scene. American games used story more literarily, particularly in what became known as "point-and-click" games, such as Sierra Entertainment's *King's Quest* and *Leisure Suit Larry*, which are "played" by moving the cursor to various points around the screen and clicking to the result of story-furthering text. These were separate attempts to provide games with a narrative foundation, and because narratives do not work without characters, a hitherto incidental focus of the video game gradually became a primary focus. With Square's RPG–*cum*–soap opera *Final Fantasy VII* in 1997, the American and Japanese styles began to converge. A smash in both countries, *Final Fantasy VII* awoke American gaming to the possibilities of narrative dynamism and the importance of relatively developed characters—no small inspiration to take from a series whose beautifully androgynous male characters often appear to be some kind of heterosexual stress test.

With that, Kohler introduced the panel's "creative visionaries": Henry LaBounta, the director of art for Electronic Arts; Michael Boon, the lead artist of Infinity Ward, creators of the *Call of Duty* games; Patrick Murphy, lead character artist for Sony Computer Entertainment, creators of the *God of War* series; and Steve Preeg, an artist at Digital Domain, a Hollywood computer animation studio. The game industry is still popularly imagined as a People's Republic of Nerds, but these men were visual representations of its diversity. LaBounta could have been (and probably was) a suburban dad. The T-shirted Boon could have passed as the bassist for Fall Out Boy. Murphy had the horn-rimmed, ineradicably disgruntled presence of a graduate student in comparative literature. As for the interloping Preeg, he would look more incandescent four nights later while accepting an Academy Award for his work on the reverse-aging drama *The Curious Case of Benjamin Button*.

LaBounta immediately admitted that "realistic humans" are "one of the most difficult things" for game designers to create. "A real challenge," he said, "is hair." Aside from convincing coifs, two things video-game characters generally need are what he called "model fidelity" (do they resemble real people?) and "motion fidelity" (do they move like real people?). Neither, he said, necessarily corresponded to straight realism. *Sesame Street's* Bert and Ernie, for example, had relatively poor model fidelity but highly convincing motion fidelity. As for the Uncanny Valley Problem, LaBounta said, "just adding polygons makes it worse. . . . The Holy Grail in video games is having a character move like an actual actor would move. We're not quite there yet." Getting there would be a matter of "putting a brain in the character of some intelligence." I was about to stand up and applaud—until he went on. One thing that routinely frustrated him, LaBounta said, was when a video-game character walks into a wall and persists, stupidly, in walking. Allowing the character to react to the wall would be the result of a "recognition mechanic," whereby the character is able to sense his surroundings with no input from the player. Of course, this would not be intelligence but *awareness*. The overall lack of video-game character awareness does lead to some singularly odd moments, such as when your character stands unfazed before the flaming remains of the jeep into which he has just launched a grenade. What that has to do with character, I was not sure. If "personality is an unbroken series of successful gestures," as Nick Carraway says in *The Great Gatsby,* the whole question of believable characters may be beyond the capacity of what most video games can or ever will be able to do—just as it was for James Gatz.

In Boon's view, when talking about believable characters, one had to specify the term. Were you talking about the character the gamer controls or other, nonplayable characters within the game? A great example of the former, Boon believed, was the bearded and bespectacled Gordon Freeman from Valve's first-person shooter *Half-Life.* Part of the game's genius, Boon said, is how Gordon is perceived. In the game's opening chapters "everyone treats you as unreliable, and you feel unreliable yourself. By the end, people treat you differently, and you feel different." Gordon's journey thus becomes your own. (Also, throughout the game, Gordon does not say a word.) As for believable nonplayable characters, Boon brought up two of the most memorable: Andrew Ryan from *BioShock* and GLaDOS from Valve's *Portal*. Both games are shooters, or neoshooters, in that *BioShock* has certain RPG elements and the "gun" one fires in *Portal* is not actually a weapon; both characters are villains. While the villains in most shooters exist only to serve as bullet magnets, Ryan (a sinister utopian dreamer) and GLaDOS (an evil computer) are of a different magnitude of invention. The gamer is denied the catharsis of shooting either; both characters, in fact, though in different ways, destroy themselves. For the vast majority of both games, Ryan is present only as a presiding force and GLaDOS only as a voice. These are characters that essentially control the world through which the gamer moves while raining down taunts upon him. In GLaDOS's case, this is done with no small amount of wit. In her affectless, robotic voice, GLaDOS attempts, whenever possible, to destroy the gamer's self-esteem and subvert all hope of survival. "GLaDOS is so entertaining," Boon said, "I enjoy spending time with her— but I also want to kill her." The death of Andrew Ryan, on the other hand, is one of the most shocking, unsettling moments in video-game history. It has such weird, dramatic richness not because of how well Andrew Ryan's hair has been rendered (not very) but because of what he is saying while he dies, which manages to take the game's themes of control and manipulation and throw them back into the gamer's face. These two characters have something else in common, which Boon did not mention: They are written well. They are funny, strange, cruel, and alive. It is also surely significant that the controlled characters in *BioShock* and

Portal are both nameless ciphers of whom almost nothing is learned. They are, instead, means of exploration.

Patrick Murphy jumped in here to say, "It's not whether the character is realistic or stylized; it's that he's authentic." In illustration he brought up Kratos of his company's own *God of War* series. Kratos is a former Spartan captain who, after being slain in combat by Ares, manages to escape Hades and declare war on the gods. Among the most amoral and brutal video-game protagonists of all time, Kratos, in Murphy's words, "doesn't just stab someone; he tears him in half. That helps sell him. Veins bulge out when he grabs things. It gives him an animal feeling that's really necessary." The narrative of the *God of War* games is set on what game designers refer to as "rails," meaning that Kratos's story is fixed and the narrative world is closed. The gamer fights through various levels, with occasional bursts of delivered narrative to indicate that the story has been furthered. It probably goes without saying that no one plays the *God of War* games to marvel at the subtlety of their storytelling, which is pitched no higher than that of a fantasy film. It is a game that one plays to feel oneself absorbed into a malignant cell of virtual savagery. Kratos's believability is served by the design and effect of the gameplay rather than the story. In short, he has to look great, which provides a fizzy sort of believability. If Kratos does not look great in purely creaturely ways, the negligible story will be dumped into the emotional equivalent of a dead-letter office. This is one of the most suspect things about the game form: A game with an involving story and poor gameplay cannot be considered a successful game, whereas a game with superb gameplay and a laughable story can see its spine bend from the weight of many accolades—and those who praise the latter game will not be wrong.

Steve Preeg, by now wearing a slightly worried expression,

opened by admitting that he was not a gamer and professed to know very little about games. But he knew a bit about believability and character. To explain the difficulty he had with animation in *The Curious Case of Benjamin Button,* he showed us "draft" shots from the digital process by which Brad Pitt's character was aged. In the earliest attempts Pitt looked undead—utterly terrifying. Just shifting the width of his eyes a tiny bit, Preeg said, made the difference between "psycho killer" and "a little boy who just got home." He showed us how he did this, and the difference was indeed apparent. Preeg then turned philosophical. In Hollywood, he said, "we have very clear goals." He worked under a director, for instance, had a clear idea of the script, and knew whether sad or happy music would be playing under the scenes he was required to digitally augment. Every eye-widening and face-aging task he was given as an animator had a compelling dramatic context attached to it, which he used to guide his animation decisions. His art was always guided. "Your characters," he said, turning to the panel, "have to be compelling in very different ways, depending on what the audience wants to do." Preeg was silent for a moment. Then he said, "You guys are going to have a very, very difficult time."

After the panel, I sought out the man who, during its EEDAR portion, turned to his colleague and said, "This can't be a good thing." His name was John Hight, and he was the director of product development for Sony Computer Entertainment, Santa Monica Studios. One of the projects he was currently overseeing was *God of War III,* a game whose budget was in the tens of millions of dollars. Yet he was no pontiff of the AAA title. Hight had also greenlit and helped fund thatgamecompany's downloadable PlayStation 3 title *Flower,* a beautiful and innovative game—a stoner classic, really—in which the player assumes control of a windblown petal and floats around, touching other flowers and gathering their petals and eventually growing into a peaceful whirling versicolor maelstrom. (When faced with releasing the tranquilizingly mellow *Flower,* no one at Sony could think of an apt category under which to market it. Hight called it a "Zen" game, and that was how it was shipped. Only later did anyone realize that the category was Hight's invention.) Hight, who was in his forties, had worked on dozens of games over the course of his career, from RPGs to shooters to flight simulators to action-adventure games. When I asked him why he had scoffed during the EEDAR presentation, he said, "The scary thing is that someone is going to enlist that data to find the ideal game that hits all the proper points, and they're going to convince themselves—and a lot of bean counters—that this is a surefire way to make money." When I asked if he could imagine any circumstances in which such data would be useful, he said, "I'm sure that our marketing people will at some point be interested, and if it helps them have the courage of my convictions, that's okay."

Hight's first title was a video-game version of the old Milton Bradley tabletop classic *Battleship* for the Philips CDI (a doomed early attempt at an all-purpose home-entertainment center that was launched in 1991 and discontinued seven years later). As the game's producer, coder, animator, and writer, Hight had no legal claim to develop *Battleship* when he began his work on it. When the time came, he simply crashed a toy expo, walked up to the Milton Bradley booth, and, after a short demonstration, strolled away with the rights. The entire game cost him $50,000. ("Very exciting," he said.) He had also been around long enough to remember an argument he had in 1994 with a colleague about whether this new genre known as the shooter was "here to stay." Hight told me, "*Doom II* had come out and done pretty well, but there really weren't many companies doing first-person shooters. I think it was seen as a novelty." Hight's colleague had insisted to him, "What more can you do? You're sort of just pointing a gun and shooting it." Hight was able to convince the man otherwise and proceed with his shooter. It became Studio 3DO's *Killing Time,* an innovative shooter for its day in terms of its gameplay (it was among the first shooters that allowed the player to crouch), setting (the 1930s), and relatively knowledgeable employment of an outside mythology (namely, Egyptology).

"When I first got into the industry," Hight said, "there were a lot of really hardcore gamers, and we were basically making games for us. We weren't making games for an audience. It was for us. And we got so specialized and so stuck in our thinking." These men's minds were typically scattered with the detritus of Tolkien, *Star Wars,* Dungeons & Dragons, *Dune*—and that was if they had any taste. Many of the first relatively developed video-game narratives were like something dreamed up by an imaginative child (a portal to Hell . . . on *Mars*! Hitler . . . as a *cyborg*!), with additions by an adult of more malign preoccupations. The writing in such games was an afterthought. For *Killing Time,* Hight told me, he "was literally writing the dialogue the day" of the recording session with the actors, one of whom approached Hight after the session was over and asked, "Do you guys ever just write a script and give it to the actors ahead of time?" A decade and a half later, Hight was still abashedly shaking his head. "Back in the day, most designers insisted writers really couldn't understand how to develop good, interactive fiction. So there was this designer–writer divide that the game industry sort of started out with."

When I asked how it could be that a panel putatively devoted to believable characters did not manage to discuss writing even once, Hight gently averred: "That panel was mostly composed of technical artists. The roots of games are in technical people. My background is computer science. I was a programmer for ten years. That's kind of how we approached it. How can we make this thing run faster? How much more can we put into the game? How can we make the characters look better?"

With its origins in the low-ceilinged monasteries of computer programming, video-game design is, in many ways, an inherently conservative medium. The first game designers had to work with a medium whose limits were preset and virtually ineradicable. There were innumerable things games simply could not do. In this sense, it is little wonder that the people who were first drawn to computers were also drawn to science-fiction and fantasy literature. As Benjamin Nugent notes in his cultural history *American Nerd,* sci-fi and fantasy literature is almost always focused "on the mechanics of the situation. A large part of the fun of reading a sci-fi series is about inputting a particular set of variables (dragon-on-dragon without magic) into a model (the Napoleonic Wars) and seeing what output you get." A video game is first and foremost a piece of software (which is why many magazines do not dignify games with italicized titles). The video-game critic Chris Dahlen, who by his own admission comes out of both a software and an "artsy-fartsy" background, argues that games "don't pose arguments, they present systems with which to interact." In this view, games are not and cannot be stories or narratives. Rather, some games choose to enable the narrative content of their system while others do not. A disproportionate number of game designers at work today come out of a systems, programming, or engineering background, which has in turn helped to shape their personalities and interests. One result of this is that it forces designers to imagine games from the outside in: *What variables do I inject into the system to create an interesting effect?*

For any artist who does not sail beneath the Jolly Roger of genre, this is an alien way to work. As someone who attempts to write what is politely known as literary fiction, I am confident in this assertion. For me, stories break the surface in the form of image or character or situation. I start with the variables, not the system. This is intended neither to ennoble my way of working nor denigrate that of the game designer; it is to acknowledge the very different formal constraints game designers have to struggle with. While I may wonder if a certain story idea will "work," this would be a differently approached and much, much less subjective question if I were a game designer. A game that does not work will, literally, not function. (There is, it should be said, another side to the game-designer mind-set: No matter how famous or well known, most designers are happy to talk about how their games failed in certain areas, and they will even explain why. Not once in my life have I encountered a writer with a blood-alcohol content below .2 willing to make a similar admission.)

When I asked Hight about these systemic origins of game design, he added that the governing systems of design have, as time has passed, become less literal and more emotional. "I think," he said, "the system's there because too many developers have failed miserably in the chaotic pursuit of something new. They have this fear of failure, so it's like, 'Okay, let's fairly quickly figure out this system.' We typically call it the 'game mechanic,' or the 'pillars' of the game. Those are our constraints, and from that we'll build around it. It's tens of millions of dollars for people to make a game. You might be the person who wasted twenty million dollars, and this is the end of your career! So you do something that's based on a proven design or proven gameplay. Why do we have so many first-person military shooters? Because it's proven those things can sell."

Video games, I told Hight, are indisputably richer than they have ever been in terms of character and narrative and emotional impact, and anyone who says otherwise has been not playing many games. Unfortunately, they began in a place of minus efficacy in all of the above, and anyone who says otherwise has probably never done anything but play games. After a kleptomaniacal decade of stealing storytelling cues from Hollywood (many games are pitched to developers in the form of so-called rip-o-matics: spliced-together film scenes that offer a rough representation of what the game's action will feel and look like), games have only begun to figure out what it is they do and how exactly they do it. Hearteningly, there seems to be some industry awareness that writing has a place in game design: One DICE presentation listed the things the industry needed to do, among them the "deeper involvement of virtual designers (and writers) into the game creative process." Alack, this banishment of the writer to the parenthetical said perhaps too much about game-industry priorities. As to whether developers could put aside their traditional indifference when it came to writing, I told Hight that I had my doubts. At nearly every DICE presentation, matters of narrative, writing, and story were discussed as though by a robot with a PhD in art semiotics from Brown. Perhaps, though, this was being too hard on the industry, which began as an engineering culture, transformed into a business, and now, like a bright millionaire turning toward poetry, had confident but uncertain aspirations toward art. The part of me that loves video games wants to forgive; the part of me that values art cannot.

Hight agreed that the audience "won't be forgiving forever," allowed that the dialogue in many games was "pretty tedious," and admitted that almost all games' artificial intelligence mechanisms delivered only half of what that term promised. But what game designers were trying to do was, he reminded me, incredibly difficult and possibly without parallel in the history of entertainment. The "weird artificial setups" of video-game narrative would begin to fade as AI improved, and already he was seeing "more emphasis" put upon writing in games. "But at the same time," Hight said finally, "our audience is saying, 'All right, what else? We're getting bored.'"

A few nights later, at DICE's twelfth annual Interactive Achievement Awards, which are the closest equivalent the industry has to the Oscars, several interesting things happened. The first was watching the stars of game design subject themselves to a red-carpet walk, most of them looking as blinkingly baffled as Zelig in the glare of the assembled press corps's klieg lights. The second was the surprisingly funny performance turned in by the show's host, Jay Mohr ("There are a lot of horny millionaire men not used to the company of women here. If you're a woman and you can't get laid tonight, hang up your vagina and apologize"). The third was the fact that Media Molecule's *LittleBigPlanet,* a game aimed largely at children, the big selling point of which is its inventive in-game tools that allow gamers to design playable levels and share them with the world, and which has no real narrative to speak of, won nearly every award it was up for, including, to the audible shock of many in the audience, Outstanding Character Performance.

The character in question is a toylike calico gremlin known as Sack Boy. There is no question that Sack Boy is adorable and that *LittleBigPlanet* is a magnificent achievement—weird and funny, with some of the most ingeniously designed levels you will find in any game—but it was also indefatigably familiar in terms of its gameplay, the most interesting feature of which is the application of real-world physics to a world inhabited by wooden giraffes, doll-like banditos, and goofily unscary ghosts. *LittleBigPlanet's* Mongolian domination of the awards became so absurd that, by show's end, Alex Evans, Media Molecule's co-founder, needed a retinue of trophy-shlepping Sherpas to hasten his exit from the stage.

The titles it bested for Console Game of the Year—*Fallout 3, Metal Gear Solid 4, Gears of War 2,* and *Grand Theft Auto IV*—were warheads of thematic grandiosity. The bewitching but more modest *LittleBigPlanet's* surfeit of awards felt like an intraindustry rebuke of everything games had spent the last decade trying to do and be—and a foreclosure of everything I wanted them to become. The video game, it suddenly felt like, had been searching for a grail that was so hard to find because it did not actually exist.

FALLOUT

HEADSHOTS

THE UNBEARABLE
LIGHTNESS OF GAMES

THE GRAMMAR OF FUN

LITTLEBIGPROBLEMS

BRAIDED

MASS EFFECTS

FAR CRIES

GRAND THEFTS

SIX

J onathan Blow has been described as "the platonic ideal of an indie game developer," but some consider him to be the industry's scourge. "I do have criticism that I level at the industry," he told me, "and if nobody else is going to say it, somebody's got to." We met in a diner in San Francisco's Mission District. Pleasingly, the most intellectually intense figure in video games today looked the part. Not his outfit, certainly—red shirt, gray slacks, blue-striped white socks, and shoes that seemed to be some sort of moccasinized cross trainer. The intensity came from his head, which sat atop shapely trapezius muscles and could have belonged to a Russian sniper: small and pitiless eyes, ghostly eyebrows, and a crew cut that seemed not the usual nuclear response to male-pattern baldness but the result of a grueling wartime imprisonment.

When I brought up his reputation, Blow waved it off. "I do a lot of things that could be seen as only helping the industry," he said, insistently. One of these was the Experimental Gameplay Workshop he leads every year at the Game Developers Conference. Another was the frequency with which he addressed conferences—and a few of these speeches have become legendary in terms of both their quality and the sharpness of their criticism. "For some reason," Blow told me, "in the past year or two, if you go to game conferences, they're under the impression that writers are good now. But if you look at what they've done, it's just clearly not the case."

Last year's GDC Montreal speech, "Conflicts in Game Design," may be Blow's finest. To listen to it is to hear the hearts of a thousand zealous game enthusiasts simultaneously implode. In the speech Blow listed the "storytelling techniques we inherently suck at," among them foreshadowing, justification, body language, and something he called "importance." By "importance," Blow meant a specific illusion of importance. One of the modern video game's most telling weaknesses is its lack of feeling for dramatic proportion—an "importance" gigantism. Why does a medium that frequently takes world-saving as its imperative so often leave one unmoved by having done so? (I know I have saved the world so many times in video games that lately I have felt a kind of resentful Republicanism creep into my game-playing mind: *Can't these fucking people take care of themselves?*) If a depiction of saving the world again and again, Blow told his audience, "is our core-value proposition, then our core-value proposition kind of sucks. It will make us money, but it will not touch people authentically and deeply." Other forms of media do not have trouble touching people authentically and deeply, and Blow set out some potential reasons as to why games are so rarely able to do this.

The first possible reason is that game designers are, by and large, unsophisticated about everything *other* than game design. "Any dumb ass can write a story for a game," Blow told his audience, "and as you have seen from playing our games, a lot of dumb asses do." A game uniquely superimposes story with what Blow calls "dynamical meaning," which is to say the meaning that grows out of exploring a game's rules and boundaries. While story can provide, in Blow's words, "interesting mental stuff" such as theme and mood, he argued that this can and should grow out of what makes games unique: play. This has the happy effect of seeming obvious once it has been pointed out, so why has this not happened? Because, Blow argued, "we don't have a culture of designers paying attention to dynamical meaning." In other words, game designers are focusing on the wrong provider of meaning, and no one is challenging them to do otherwise.

The second possible reason is that the video-game form is incompatible with traditional concepts of narrative. Stories are about time passing and narrative progression. Games are about challenge, which frustrates the passing of time and impedes narrative progression. The story force wants to go forward and the "friction force" of challenge tries to hold story back. This is the conflict at the heart of the narrative game, one that game designers have thus far imperfectly addressed by making story the reward of a successfully met challenge. According to Blow, this method is "unsound," because story and challenge "have a structural conflict that's so deeply ingrained, it's impossible" to make game stories strong. Can better writing solve this? In Blow's mind, it cannot. The nature of the medium itself "prevents the stories from being good."

A good game attracts you with melodrama and hypnotizes you with elegant gameplay. In effect, this turns you into a galley slave who enjoys rowing. For Blow, this approach to game design does not satisfactorily address game interactivity. If story is a stately "presentation of events," interactivity "directly sabotages . . . the way that effective stories are told." According to Blow, games tell us much we do not want or need because they lack the authorial filter of traditional storytelling. Mature mediums deal with such formal problems as a matter of course. A stage play can use blatantly unconvincing scenery and a film can shoot a night scene implausibly awash with light because both mediums have figured what is formally distracting for an audience and what is not, what an audience needs and what it does not. Games are a long way from that confidence, and as a result their stories have, at best, biplane functionality.

As Blow reminded his audience, many—indeed, *most*—forms of creative expression have no truck with story. (Even forms of creative expression that do include story use it in a way that leaves no doubt that the real art is happening elsewhere, as in, say, opera.) Blow attributes the video game's umbilical attachment to story to the influence of film. This fatal attraction has caused games to lack what Blow calls "the clarity of consequences." If a work of art's conceptual underpinnings are "fake, unimportant, arbitrary, and careless," it cannot be profound or important or have deep meaning to people. "We have adopted design practices and ways of making games," Blow told his audience, "that *are* fake, unimportant, arbitrary, and careless." Blow believes that everyone who plays games can sense these conflicts whether he or she knows it, and they short-circuit every game's emotional appeal to its audience.

Blow had been on these ideas for several years, but only recently had he been able to find a broad platform for them. Born in 1971, Blow had, by his own admission, worked "nowhere very consistently" for most of his career. After studying computer science and literature at Berkeley—from which he neglected to graduate—he drifted, writing the occasional short story, working the occasional for-hire job in the tech industry, and thinking about video games. The work he did do was almost always as "a consultant and outsider kind of guy" tasked with what he called "hard problems." None of these gigs lasted more than a few months.

"I have this weird thing," Blow told me, "where my motivation will just totally flag if I feel like I'm not doing something important." By twenty-three he had saved up $24,000 ("This is back when $24,000 was real money!") and with a university friend concocted a grand plan. "He was getting out of grad school, and we were like, 'Let's start a game company!' We didn't know anything about games. We just started doing it." The dream lasted for four tractionless years. In the meantime Blow had begun writing a column for *Game Developer* magazine and put himself forward to lead the Experimental Gameplay Workshop at GDC. Throughout this period he started many games but never finished any of them. He also had an office in Oakland, which he shared with some friends, who were busy not finishing any of their games. A few years later, he told me, "I had this really strong idea. As soon as I started doing it I was like, 'Yeah, this is different from what I was doing. And I see that I can do a lot of things that are different, in different ways.'" For the first time in his professional life, Blow could honestly tell himself, "I care about what I'm doing, and I'm doing things the industry is not going to do if I don't do them."

Many game designers discuss their work fully aware they have leased their souls to one devil or another and almost manage to convince you that their capitulation, however regrettable, was necessary. Talking to Blow was not like this. For years, Blow had been accused of being too idealistic and possessing interesting ideas that had no commercial application. While working on his "strong idea," he told me, and the game it eventually became, many people "said things to me about what I should do to make it sell, but I said no. That's the common wisdom: You can't just *make* the thing." The thing he finally made was a downloadable game for the Xbox 360 called *Braid,* into which he sank $200,000 of his own and borrowed money. Blow created *Braid* in open defiance of many commercial orthodoxies—and it made him wealthy enough that, when I asked for some ballpark idea of how well the game had done, he requested that I turn off my tape recorder.

Big, dumb, loud action games can be highly sophisticated as games, though their stories—the thing they are trying to use as vehicles for meaning—probably will not be. The so-called art game, of which Blow is now a leading proponent and *Braid* a source text, has risen up in response to this. Many art games are abstract or purposefully old school. They work off a few basic assumptions: Games have rules, rules have meaning, and gameplay is the process by which those rules are tested and explored. In many art games, it is gameplay and not story that serves as the vehicle for meaning.

The language of gameplay is driven by sensation rather than words. Like music, it can have themes and motifs, however distantly apprehended. What Blow did with *Braid* was, yes, braid gameplay with themes and motifs. The theme in question is time, which the gameplay forces the gamer to literally and conceptually play with and subvert. Unusually, *Braid* does this in the form of the platformer, a venerable but severely limited genre, even though the platformer is what many nongamers imagine games to be, largely because a number of the form's most famous characters were raised in its nursery: Donkey Kong, Mario, Sonic the Hedgehog. Founded upon running along planes, climbing ladders, leaping over enemies and across chasms, the platformer is among the most childlike genres in that it provides worlds stylized for pure play, but many platformers are known for their fearsome difficulty. Mastering a platformer such as *Donkey Kong* is not play; it is a psychically crushing process of memorization and reflex mastery.

Those who imagine all video games to be a variation on the platformer formula are, in some ways, more correct than not.

Conceptually speaking, the platformer may be the most arche-
typal video-game genre. A role-playing video game takes its core
inspiration from tabletop games such as Dungeons & Dragons,
while the first- and third-person viewpoint of many other games
comes straight from the language of film. A platformer, on the
other hand, has very few traceable antecedents, and those it does
have—the static, sideways storytelling of Egyptian hieroglyphics,
say—feel very distant indeed. *Donkey Kong* and *Super Mario Bros.*
are designed with ant-farm intricacy, and the objects that govern their worlds—cheerful industrial jetsam such as impractically
tiny elevators and glowingly magical hammers; Venus-flytrap-
inhabited pipes and small sinister turtles fishing off clouds—have
an overwhelming aura of not being able to exist elsewhere, in any
other world, real or imagined. The platformer world is one of
bright, dynamic, interrelated flatnesses, and when I am playing a
great platformer I sometimes feel as though I am making my way
through some strange, nonverbal poem.

Like a poem, a great platformer does not disguise the fact that it
is designed, contains things you cannot immediately see, and
rewards those willing to return to them again and again. One of
the greatest and strangest platformers in this respect is Nintendo's
Metroid, which stymied a friend and me for weeks when we were
boys. (Some may object to calling *Metroid* a platformer, despite its
many platforming elements. I would argue that *Metroid* is, in fact,
the first open-world, nonlinear platformer.) My friend and I duti-
fully explored *Metroid's* every interplanetary cranny until, finally,
there seemed nowhere else to go. The gameworld simply ran out,
and, needless to say, we had no Internet to turn to. One day my
friend and I were playing *Metroid* in a desultory, pointless way,
rolling ourselves into a morph ball and laying bundles of explosives charges because we liked the way the bombs launched us

harmlessly into the air. But we made a strange discovery when, in an obscure part of the *Metroid* world, our bombs went off and part of the floor disappeared. This revealed a secret chute through which to fall and an entirely new part of the gameworld to explore. My friend and I were so happy we embraced. (Actually, I may have cried.)

Because no other genre is quite so content to risk gamer frustration as the platformer, no other genre provides quite the same feeling of satisfaction when that frustration is overcome. I know that when I play superb modern platformers such as Nintendo's *Super Mario Galaxy* or *LittleBigPlanet* with my nieces, and work with them to solve puzzles or figure out ways around seemingly insurmountable spatial puzzles, the joy that comes with having done so has nothing to do with story or character or dramatic meaning but rather feeling your mind identify some mystical, vaguely mathematical outline. Platformers like *Super Mario Galaxy* and *LittleBig-Planet* make the world feel newly, complicatedly strange but also conquerable, and thus remind me of what was actually *fun* about being a child.

In designing *Braid,* Blow opted for the platformer to evoke this very "childhood" feeling. A platformer, he told me, "because it has this simplicity, was the simplest kind of world that I could think of that had a small number of rules and a small number of interactions, and where you, as a player, could have a pretty good idea of what's going to happen a few seconds into the future. I wanted to take that and make it not simple."

*Braid'*s player-controlled character is a young unnamed boy, though Blow refers to him as "the dude." With his floppy hair and vaguely Etonian schoolboy tie, the dude looks a bit like Hugh Grant in bantam cartoon form. He is also searching for a princess, and *Braid* begins with the reliable conceit of opening a door into another world. The first thing to be said is that the world of *Braid* (which was created by the gifted artist David Hellman) is beautifully aglow—an arcadia of chuggingly locomotive clouds, heartbreaking dusk, small scurrying creatures, and lustrous flora. The second thing to be said is that it sounds like no other game, and certainly no other platformers, which typically plug into the gamer a sonic IV of bouncily reassuring and sometimes tormentingly repetitive music. *Braid*'s sound track—which was licensed rather than written explicitly for the game—is slow, string-heavy, and celestially lovely. Half of the pleasure of *Braid*, at least initially, is simply to stand there, look, and listen. The combination of the visually beautiful and the music's plangent lushness is part of what makes *Braid* look so happy but feel so sad. This was, Blow told me, purposeful. He wanted the gamer to think, *"This isn't as happy a place as I thought or hoped it would be."*

In many ways, a video game can be viewed as a pure text in the same manner one views a film or work of literature. There is, however, at least one important difference. Films and works of literature are composed of signs and signifiers that share some basic similarities with their counterparts in the observable world. In many early video games (*Tempest, Pac-Man*), the signs and signifiers are rarely connectable to the observable world in any rational way. Only God or possibly Timothy Leary knows what *Tempest*'s wall-crawling light spiders are supposed to depict. But it does not matter. All that matters is what the light spiders of *Tempest* signify within the world of *Tempest*. By this analysis, early video games such as *Pong* and *Spacewar!* are, developmentally speaking, cave paintings, whereas *Tempest* and *Pac-Man* are something like modernism, albeit a modernism of necessity. Within the evolution of video games, no naturalistic stage between the primitivism of *Pong* and the modernism of *Tempest* was possible due to the technological limitations to which game designers were subject. When naturalism did come to video games in the early 1990s—the enabling of true, in-game, three-dimensional movement was as climacteric a development for the medium as the discovery of perspective was for painting—it was so breathtaking that many forgot that naturalism is not the pinnacle but rather a stage of representation. With *Braid,* a considered, impressionistic subversion of the "realistic" has at last arrived, and Blow may be as spiritually close to a Seurat or Monet as the form is likely to get.

Braid does not allow the dude to die. Instead, when the dude runs afoul of one of the gameworld's strange little creatures or is beaned by a cannonball or falls into a molten fire pit, the game simply stops. You then rewind the game to a point of safety and try again. What initially feels like a clever gimmick (one that, admittedly, other games had employed before) eventually comes to have considerable emotional power: The dude is searching for someone he lost but who may not be recoverable, even with the subvention of time travel. Many of *Braid's* puzzles require the gamer to experiment with this time-travel mechanic—dropping into a hole to retrieve a key and rewinding to be sucked back up to the precipice with the key safely in hand—and later levels present some truly fascinating temporal riddles. In one example, rushing the dude forward results in backward time travel for everything else on the screen (including the music), much of it into a place that appears to prevent further exploration. During such moments *Braid* becomes a moving spatial crossword puzzle—a game in four dimensions.

Braid is also implausibly difficult. A friend and I completed it, but only with the aid of chronic YouTube consultations. In his GDC speech, Blow argued that challenge is, too often, squandered in games, too many of which hold out "faux challenges" and over-reward the player for having surmounted them. "A game doesn't need to be difficult," Blow has said, "it just has to be interesting. It has to convince the player that their actions matter." This is among the most compelling achievements of *Braid*. While I occasionally despaired of my ability to solve certain puzzles, the game never frustrated me. Its difficulty is interesting because it is not arbitrarily difficult. It is meaningfully difficult, because, again, it forces you to think about what subverting time really means and does—and what it cannot mean, and cannot do.

Blow filled the world of *Braid* with scaffolds of sneaky autobiography, which may be what provide it with its unusual melancholy and corresponding emotional significance. It feels as though the person who created it was trying to communicate something, however nameless and complicated. It feels like a statement, and an admission. It feels, in other words, a lot like art. While Blow disputes the oft-floated claim that the game was his response to a breakup, the time control had its origins in an abandoned billiards game of Blow's design in which the player was able to see exactly where his or her shots would come to rest. It was a simple idea—so simple, in fact, that Blow found it unworkable—but it was his fascination with foreseen consequence, born of a mind sick of failure, that inspired the backward march to the platformer. Once again, the game's meaning recombines: For Blow, creating *Braid* was aesthetic time travel.

Another convention of the platformer is the ability to jump on and bounce off enemy characters, sending them tumbling into the offscreen afterlife. In most platformers, a successful landing upon an enemy results in a happy *boing* of victory. The creatures of *Braid,* however, make a disappointed, almost booing sound. "That guy didn't want to get jumped on," Blow told me when I asked about this, and while the enemy creatures of *Braid* are, in his words, "certainly subhuman," Blow insisted on giving them expressive, vaguely human faces. "I wanted it to feel like, 'Yes, there are things you are supposed to be doing, but they have consequences.'" In Blow's analysis, most video games are "all about *not* introducing doubt about what you're doing: 'Hey, the enemy soldiers captured the hostages, and I'm running up and shooting the enemy soldiers and I'm rescuing the hostages.'" Blow referred to this style of gameplay as one that puts the player in an "animal reaction mode," which "can't matter to me on an intellectual, emotional level the way a lot of good art does."

When I asked Blow if he was categorically opposed to games that involved gratuitous amounts of combat, he surprised me by saying that he admired many things about *Grand Theft Auto III* and *Gears of War.* "But," he said, "I am against the entire industry making only that. When we only make that, what does that mean about us and our ability to approach subjects about humanity on the whole?"

One of the bugbears of the sharper video-game blogs is why cultural validity and respect persist in eluding the reach of the video game. This question tends to bedevil gamers rather than game designers, most of whom it is difficult to imagine sitting down in a game's planning stage and asking themselves if what they are making will be art. I do not fault gamers for asking the question; all of us want the reassurance that we are not spending absurd amounts of time on something without merit. I asked Blow what he thought about the question. "It's a prerequisite," he said, "that to be respected as somebody who is saying important things, you have to have important things to say. We're not really trying to have important things to say right now. Or even interesting things to say. People want to have an interesting story, but what they mean by that is this weird thing that comes out of copying these industrial Hollywood processes. The game developer's idea of a great story is copying an action story." He shook his head. "Isn't it a little obvious that that's never going to go anywhere?"

FALLOUT

HEADSHOTS

THE UNBEARABLE
LIGHTNESS OF GAMES

THE GRAMMAR OF FUN

LITTLEBIGPROBLEMS

BRAIDED

MASS EFFECTS

FAR CRIES

GRAND THEFTS

SEVEN

Godforsaken is often used to describe the world's woebegone landscapes. But to say that God has forsaken something, there must be some corresponding indication that God had ever shown any interest in it, and, in the case of Edmonton, Alberta, this was not immediately apparent. On the evening of my arrival, at least, the temperature was close to the magic intersection at which Celsius and Fahrenheit achieve subzero parity. I was in Edmonton to see Drew Karpyshyn, the head writer of BioWare's *Mass Effect,* a science-fiction role-playing game that some have held up as one of the best-written console video games of all time.

There is a nontrivial divide separating the relative achievements of console and PC games in any number of areas, but how "well written" console games are when compared with PC games, which have historically been more text-heavy, is especially contentious. Among the PC gamers of my acquaintance, Black Isle Studios' RPG *Planescape: Torment* is often cited as being more thought provoking and literarily satisfying than any console game. In this respect, BioWare's console games have an instantly recognizable style: that of seeming like PC games (a famously persnickety and piracy-plagued market that BioWare, unlike many developers, has not abandoned). What distinguishes the BioWare style is an unshakable reliance on dialogue and narratives with all manner of bureaucratic complication. What also distinguishes the BioWare style is gameplay longevity: I have had moderately meaningful relationships in which I invested less time than what I have spent on some BioWare games.

All of BioWare's titles have been RPGs of one stripe or another, with an early concentration on the dungeon fantasia, an RPG subgenre that is extremely difficult to do well and virtually impossible to sell beyond its niche audience. The first BioWare title to move beyond the cleric-and-dwarf sodality was 2003's *Star Wars: Knights of the Old Republic* (known, in vaguely Neanderthal vernacular, as *KotOR*), which is set four thousand years before the events lamentably depicted in *The Phantom Menace*. With *KotOR*, BioWare was in danger of simply swapping one shut-in constituency for another, but it was a game of such narrative superbity that even non–*Star Wars* fans took note. While the care the game lavished on the *Star Wars* universe was considerable, the way *KotOR* handled dialogue indicated a solidifying methodology. Here, as in the later *Mass Effect*, almost every conversation and encounter initiated by the gamer can lead to multiple and often drastically different outcomes, some of which bring you in line with the Force, some of which tempt you down the path of the dark side of the Force. The game changes—as does your character's appearance—depending on where he or she falls along a spectrum of in-game morality. Although open-ended conversation may sound like a relatively simple game mechanic, when it is done well that is most assuredly not the case. The technology BioWare uses to manage in-game dialogue is closely minded, and parts of it are patented. No one, then, makes more conversationally driven console games than BioWare. When these games are played in proper solitude, the marathonic dialogue rarely becomes an issue. To watch someone *else* play a BioWare game, however, is to ponder the boredom-curing upshot of punching oneself in the face.

For gamers with dreamier turns of mind, the somewhat draggy narratives of games centered upon the unpredictability of conversation and encounter provides half of the enjoyment. The dynamism of combat or movement has never been the strength of the RPG and never will be. Indeed, the notion that involved narrative has any place in video games at all begins in the RPG—a fact I have heard more than a few game designers lament. While most game genres have ransacked the devices of film, the RPG has in many ways drawn from the well of the literary. This is the source of many game designers' suspicions. Why construct an entire genre upon the very foundations (character, plot, theme) that have given games such trouble?

The man I had been seated next to on the plane into Edmonton, a *KotOR* fan from way back, underwent a spontaneous volubility transfusion when I explained the purpose of my visit. The woman manning the booth at immigration control gave my passport a hard, prideful stamp when I revealed the name of the local company I would soon be seeing. My Lebanese cabdriver, while making his way along an icy highway at speeds approaching fifteen miles an hour, nearly lost control of his sedan when I revealed my destination. "Big company," he said. "Powerful company!" He then asked if BioWare's physician founders, Ray Muzyka and Greg Zeschuk, were still practicing medicine. (BioWare's name derives from its origins, long jettisoned, as a medical technology company.) I didn't think so, I said. "That's very sad," the driver said. "The world needs doctors." Apparently, he was not a *KotOR* fan.

BioWare's offices take up the second, third, fourth, and fifth floors of a tiered, charmlessly concrete office building on the south side of Edmonton. Once inside, I submitted to the required processes of game-studio entry: shaking the hand of the extremely pleasant PR person (who would be sitting in on my meeting with Karpyshyn to monitor "the messaging he's putting out"), scrawling my name across the nondisclosure agreement, gladly agreeing to a quick tour.

At my request, the tour momentarily paused when we came to eight tall cabinets filled to capacity with books and old board games. That BioWare would have a large library was not surprising: Its games are noted for the vastidity of their worlds, all of which must be designed and populated and inhabited. Along with all the expected stuff (pop novels like *Jurassic Park*, old Dungeons & Dragons reference guides, an inordinate number of books whose titles included either *realm* or *lance*), BioWare's library went beyond Advanced Nerd Studies: *The Ultimate Book of Dinosaurs; Architecture of North America; Giants of the Sea; Chinese Grammar; Guns, Germs, and Steel; The Celtic Book of the Dead;* and *The Complete Idiot's Guide to World Religions*. At this last I looked over at my handler. "We have to come up with a lot of lore," she said with a shrug.

In the common area, five youthful BioWare employees were gathered around a massive flatscreen television and playing *Street Fighter IV,* which had just been released for the Xbox 360. That is, two were playing while the other three watched. From their total emotionlessness I gathered that this somehow qualified as working. Although none of these young men appeared old enough to rent a car, I was told that the average age of BioWare's employees was, in fact, thirty. Compared with the rest of the industry, this practically qualified BioWare as an assisted-living provider.

The tour concluded with a meeting room that illustrated the degree of fan loyalty inspired by BioWare's games. Hung on the walls of this room were nineteen painstakingly detailed woodcut plaques that bore the design-specific titles of every game BioWare has released. The artist responsible had sent the woodcuts to BioWare at no cost in order to show his "appreciation for years of great gaming." I tried to imagine someone, anyone, doing such a thing for Paramount or Random House. This was quite impossible.

Drew Karpyshyn was a large, tree-trunk-solid man, his buzzed-down hair bringing to mind a soldier a few years clear of active duty. His face, however, bore few traces of its thirty-seven years, and I wondered if there was something about a lifelong commitment to sci-fi and fantasy (Karpyshyn is also a science-fiction novelist) that kept one boyish. As we sat down, I told Karpyshyn that, having now visited Edmonton, I believed I understood why BioWare made such long, involved, complicated games. He laughed and admitted that there was something to this. "There's a huge amount of video-game talent in Edmonton," he said. "Sixteen hours of darkness? What else are you going to do but play games?" Born and raised in Edmonton himself (though his ancestral heritage is Ukrainian), Karpyshyn had recently decided to relocate to BioWare's Austin, Texas, office. "I'm done with Canadian winters."

When I confessed to having spent around eighty hours playing *Mass Effect,* I could tell from Karpyshyn's pop-eyed reaction that even he considered this excessive. Among video-game genres, only the RPG was capable of subjecting me to such a lengthy enslavement. What was it, I asked him, about the RPG? If play is freedom with structure, do not rule-bound genres like the RPG simply add another and possibly unnecessary layer of structure over the structure of video-game play? Why do so many people respond to that? I certainly respond to it, I told him, but I was not always certain I wanted to.

"Story has been more important in RPGs than it has been in other types of games," he said. "That's one thing that appeals to me, as a writer. Now that's starting to change. You're seeing story propagate across the different genres. A lot of games out there have a very interesting story, but it doesn't really matter what you do. With RPGs, the fact that you can actually influence the story, and control it in some way, and have a different experience—a personal and individualized experience—is very important."

Even more significant, he told me, was the RPG's defining characteristic: the player's ability to create his or her own character. In *Mass Effect* you are presented with a name: Shepard. Almost everything else is open to alteration. *Mass Effect*'s catalog of physical features is not as large or varied as some games with character-creation options (you have your pick of a dozen or so noses, two dozen hairstyles, an array of facial scars, and so on), but the game provides an additional interesting measure of psychological customization. Shepard can be the sole survivor of a long-ago massacre, a storied war hero, an erstwhile criminal, and so on. (The specific past and hang-ups with which your Shepard is saddled will be reflected within the game's narrative and often determine the kind of people you will meet in the gameworld.) "I don't really identify with a premade character," Karpyshyn said. "When I make a character—even if I don't make the character look like me—that is the character I'm inhabiting through the game. Even if it's a female character or not even a human character—it doesn't matter. I feel connected."

I chose not to get into how long I took in designing my Shepard. The fruit of my labor was a striking green-eyed redhead with drill-resistant cheekbones and nicely plumped lips. Long after I finished *Mass Effect,* I consulted YouTube to rewatch a few of its key scenes and was confronted by a series of rank imposters: bald Shepards, Asian Shepards, blond Shepards, black Shepards, and (most appalling) *male* Shepards. This was a form of video-game interactivity that slid around the criticism of Jonathan Blow: It was an *imaginative* interactivity that in many ways resembled the reading experience, in which characters are cast and costumed in the mind's definitive privacy. An RPG such as *Mass Effect* literalizes this process. The YouTube Shepards struck me as imposters because that is what they were.

The special resonance of the created character will amount to very little if the story she becomes part of is badly or lazily conceived. Because the typical RPG tells its story through serial conversation, dialogue is where the genre lives. More frequently, it is also where the genre dies. Many RPG characters have a peerless gift for antispeech, from the lobotomized Shakespeare of the average fantasy game to the exotically inane nomenclature of the average sci-fi game. No other genres tip so easily into silliness when trying to be deadly serious, and there is no purer indictment than that. In light of this, I had devised a simple scenario: If I am playing an RPG, and the characters are talking, and my response to a woman of any foreseeable nudity walking into the room is to instantly turn the game off, I know that what I am playing does not have much adult nourishment. *Mass Effect* almost always passed this test. When I asked Karpyshyn about the unusual facility his games had with dialogue, he said it was attributable to the fact that BioWare simply has more writers than most developers: thirteen in its Edmonton office and almost as many in Austin. One happy result of this was the quality control of competitive evaluation. "You *know*," he said, "that your stuff is going to be seen by other writers." As far as he was aware, no other developer had as many writers on staff.

BioWare's writers, as full-time employees, are involved in the creative process from beginning to end. "A lot of companies," Karpyshyn told me, "will bring writers in at the beginning and say, 'Do an outline,' or bring them in at the end and say, 'Write a script.'" While the game industry was full of what Karpyshyn called "nightmare stories" of writers being abused, ignored, and discarded by developers, "BioWare respects the writing process." BioWare also indulged the writing process: The script for *Mass Effect* contains three hundred thousand words.

Despite science fiction's sui generis presumptions, most sci-fi worlds—imagined at the balance point of the evolutionary and the point-mutational, the cautionary and the aspirational—are imitative. Bad science fiction often seems to have not enough influences or too many obvious influences. Part of what made the world of *Star Wars* so attractive was its odd ingredients: Arthurian legend, the samurai film, the Western, World War II dogfight footage, Nazi propaganda films. George Lucas's vision was as imitative as any other, but what it imitated was, at the time, crisply eccentric.

In 2005 a small group of BioWare writers, designers, and artists sat down at the conference room table Karpyshyn and I now shared. George Lucas was very much on their minds. The team had one goal: to create a "massive science-fiction game" that took place not a long time ago, in a licensing agreement far, far away, but in a universe of BioWare's creation. (Self-generated fictional universes are referred to as "intellectual property," or IP. The way most developers throw the phrase around leaves little doubt as to which half of it is more coveted.) They knew the game would be done in what Karpyshyn called "the BioWare way—very story-heavy," but that was all they knew. Was it far future or near future? Would it be darker sci-fi, as in *Blade Runner,* or something more optimistic, as in *Star Trek*? For several weeks they talked about their favorite science-fiction films and novels. All had elements of special intensity, but what made these elements so affecting, and why? A list was made, and they realized that what these elements shared was not that they looked great or sounded cool, which is the point at which many works of sci-fi kick back and call it a day. Rather, these elements tapped into the emotions to which science fiction has privileged access: hope for and wonder at the potential of human ingenuity and, of course, fear of the very same. Rather than mimic the particular sci-fi elements that gave rise to these emotions, the emotions themselves became BioWare's goal. "I think," Karpyshyn told me, "that this is the step that a lot of games miss." When I asked him if the list was still extant, he said that it was—and under no circumstances save for imminent Armageddon could he show it to me. *Mass Effect,* the first game of a projected trilogy, had scratched the list's surface with no more than a pinkie fingernail.

I believed I could identify one item on the list without seeing it. A few characters in *Mass Effect* have an ability with something called "biotics," defined by the *Mass Effect* Wikipedia page (which is a mere one hundred words shorter than that of President James Earl Carter) as "powers accessed by the characters using implants that enhance natural abilities to manipulate dark energy." In gameplay terms, this amounts to the very enjoyable ability of throwing enemies around the room, sabotaging their shields, levitating them into positions of extreme vulnerability, and clobbering them with invisible freight trains of directed energy. One did not need any Dagobah training, I told Karpyshyn, to regard biotics as a step beyond homage. To his credit, Karpyshyn laughed. "I can see why people would say there are similarities, but let's be honest: The idea that you can, with your mind, influence the world around you in miraculous ways is not a new idea." Karpyshyn made sure to point out that, unlike the Force, which is one of the least-thought-through aspects of the *Star Wars* universe (why, if Darth Vader can choke at will, does he even bother with a lightsaber?), the use of biotics was governed by internally consistent rules that went beyond the expected gameplay mechanic of waiting for one's biotic powers to "recharge" after overuse. He claimed these rules were a product of BioWare's "sciencey" culture.

As well written as it is, *Mass Effect* neither fails nor succeeds on literary terms, for no game could. Literary sci-fi, in fact, has an overriding advantage: its invulnerability to visual disappointment. Those who are inclined to cherish the patently unconvincing have, thanks to a century of science-fiction films, one of the widest selections in history. Even beloved works of sci-fi have a disgraceful ratio of arresting aliens to hideously inadequate aliens, which made J. G. Ballard's dismissal of the *Star Wars* cantina scene ("Muppets in space") so devastating. It is well to remember that science fiction is not a license for speculative biology: If a real alien is ever discovered, it will probably not look like anything we can imagine. Admittedly, then, a "convincing" alien is subjective. Science fiction may err when it imagines the alien as nothing more than a ridged forehead and a surly mood, but arguing why a Klingon is inferior to, say, E.T. will reveal no clear path to victory. About all one can argue is that E.T. was more rigorously imagined than a Klingon.

Happily, the aliens of *Mass Effect* are E.T.'s, not Klingons. The face of a turian, for instance, somehow resembles a cross between a camel and an artichoke. The discomfortingly sexy asari look like what might have happened if Veronica Lake, the Blue Man Group, and a hood ornament had a child. Krogans comport themselves like large, reptilian, extremely pissed-off elderly retirees. The synthetic geth, which a large part of the game is spent killing on sight, look like nothing so much as incredibly evil desk lamps. However ridiculous the above comparisons make these aliens sound, they are all as mysteriously evolved but pleasingly convincing as the snaggled and antennaed denizens of the deepest parts of the sea.

No matter how wild the speciation of *Mass Effect* gets, its world has the imaginative solidity of wrought iron. As Jesse Schell writes in *The Art of Game Design,* the illusion of internal consistency in video games is as important as it is frail: "[U]nlike story-based entertainment, where the story world exists only in the guest's imagination, interactive entertainment creates significant overlap between perception and imagination, allowing the guest to directly manipulate and change the story world. This is why videogames can present events with little inherent interest or poetry, but still be compelling."

The other burden placed on *Mass Effect* is its need for quality voice acting. Fortunately, not a single performance in the game is less than competent, and several are startlingly good. While the film actors corralled by *Mass Effect*—Seth Green and Keith David among them—perform ably, the game's most glorious performance is that of Jennifer Hale, now widely viewed as the Olivier of video games. Hale plays Shepard—if, that is, one opts for a female Shepard. When I asked about Hale, Karpyshyn said, "She's brilliant. Her performance is so powerful, but it still allows you to feel like you are the character. It doesn't distance you, and that's very hard to do." Hale's performance is even more impressive given the constraints she was under. In all of BioWare's previous games, as in most RPGs, the controlled character "speaks" when the gamer selects a desired statement or response from a proscribed menu, with the words themselves going unheard. Because *Mass Effect* is fully voice-acted, the game's designers had to concoct a mechanic that would prevent the gamer from being twice exposed to what he or she wished to say. (Plus, Karpyshyn said, the actor would "never quite say it the way you said it in your head.") The solution to this problem is what BioWare calls the Paraphrase System. When, at one point in the game, Shepard is sold out by the loathsome careerist Ambassador Udina, the Paraphrase System provides these responses: "This is a mistake," "This is stupid," and "You son of a bitch!" If one picks the first two choices, Shepard's response is fairly tepid. If one picks the third choice (and I certainly did), Shepard snarls, "Nobody stabs me in the back, Udina. Nobody." The acting challenge here is obvious. While performing the game's many hundreds of exchanges, Hale had to express the spirit of the revealed paraphrases but remain tonally neutral enough to allow different conversational paths to lead to and depart from what Shepard is saying. At the same time, the nonlinear nature of the Paraphrase System prevented Hale from being able to perform any one thread of conversation all the way through, which would have been impossible in any event, as the game script bore no resemblance to that of a film. In effect, Hale was asked to provide the branches of a tree she could not even see. Karpyshyn noted that, for an actor, the only equivalent experience would be performing several different takes of a scene simultaneously.

The gratitude with which Karpyshyn spoke about Hale's performance suggested that the collaborative nature of video games was the source of frustration as often as not. When I asked him about this, he admitted, "With a collaborative medium it's much easier to get bad art. Games have gotten so complex that you need this huge group of very talented people. *Mass Effect* had a team of about one hundred and twenty. With games, you take a lot of pride in saying, 'I was part of this great team.' " He wrote novels as a partial tonic to this: "When the novel comes out, I can say, 'That's mine. I made that.' "

Later I read Karpyshyn's novel, which is set in the *Mass Effect* universe. Although a perfectly cromulent science-fiction novel, *Mass Effect: Revelation* pleased me perhaps one-hundredth as much as *Mass Effect* the game. Was this because the game was interactive and the novel was not? Most assuredly, no. Novels are vigorously interactive, and video-game interactivity (the limitations upon which are legion) is frequently overstated.

The meaning created through reader–writer interaction is categorically different from the meaning created through gamer–game interaction. The way a reader reads a novel may change; the way a writer understands her novel may change; but the novel itself remains invariant. I could debate the meaning of Karpyshyn's *Mass Effect* novel, but a debate over the meaning of *Mass Effect* the game would be comparative, not interpretive. What did you do on Noveria? I decided to extirpate the rachni species. What about the wicked Dr. Saleon? I gunned down the defenseless cretin in cold blood. What about the poor enemy scientist cowering behind her desk on Virmire? When she begged for mercy, I uncharacteristically decided to grant it—the whole Dr. Saleon episode had left a bad moral aftertaste. Armor? By game's end I was wearing Scorpion VI armor (updated with Medical Interface V). Weapon of choice? The Tsunami VII assault rifle (armed with Hammerhead rounds and tricked out with a barrel-lengthening rail extension). The world of *Mass Effect* was conceived with the gamer in mind, and the shadow of this final collaborator falls distractingly across every page of *Mass Effect: Revelation*. This was an issue even Karpyshyn had anticipated. The reason there is no Shepard in his *Mass Effect* novel, he explained, is because Shepard was not his character but mine.

There are two important things I have not yet addressed about *Mass Effect*. The first is its narrative. The game opens nineteen years after human beings made first contact with an alien federation overseen by the Council, which calls a spacefaring dreadnought known as the Citadel home. Humans are gaining influence within various Council organizations (though not the Council itself), and this is very much to the irritation of many other races. (The degree to which you can inflame or dampen alien racism among your own crew is one of *Mass Effect's* most interesting quandaries. I almost always inflamed it—and felt enjoyably bad in doing so.) One of these Council organizations is a paramilitary police group known as the Spectres, of which Shepard, at the end of the game's opening act, becomes the first and only human member. The title, meanwhile, refers to technology left behind by an ancient, vanished culture that allows the many races of the galaxy the luxury of intergalactic travel. When it becomes clear that Mass Effect technology has another, far more sinister application, you must stop a rogue fellow Spectre named Saren from enabling it. Along the way one notices the clever wainscoting that all good sci-fi specializes in: The human discovery of Mass Effect technology took place on, cleverly, Mars; the first outer-edge human space station is named in honor of Yuri Gagarin, and one nebular star cluster bears the designation *Armstrong;* an overheard newscast describes an alien species mounting a performance of *Hamlet* that will use pheromones in place of dialogue; and so on.

All of this lines the corridors of Mass Effect with panels of preexistence, and with the illusion established that much has happened here, one believes that much else can.

The other thing I have not yet addressed is Mass Effect's gameplay. In some ways, Mass Effect is not a very good game, at least not according to the criteria by which most games are judged. It is designed to operate as a third-person shooter, most examples of which, following the success of Gears of War, are obliged to provide a cover mechanic. In most games, cover is achieved by pressing the assigned button when one's character nears a protective object. This allows one to pop over or lean around the cover-providing object and spray enemy positions between volleys of return fire. A cover system works when it does not feel too sticky. Cover should be easily attained and just as easily abandoned. The intuitive and responsive cover mechanic in Gears, though not perfect, is still the industry standard. The fighting in Mass Effect has an enjoyable briskness, with enemy bodies disintegrating upon the fatal shot, yet its cover system is sometimes unresponsive. While not hard to get into, cover is often difficult to break out of. When one's shields are down to a final power cell and one's enemies have sneaked into a flanking position, finding oneself stuck in cover and unable to respond defines gameplay frustration. Other mechanical issues are even more puzzling. For reasons known only to BioWare, grenade-throwing has been assigned to the Xbox 360 controller's unassuming "back" button, to which few games cede any gameplay function at all. Indeed, throwing a grenade at a platoon of geth with the "back" button feels as fundamentally mistaken as using the volume knob on your car stereo to roll down the driver's-side window.

Because Mass Effect is a role-playing game, at least 20 percent of one's playing time is spent in various menu screens allocating talent points and upgrading weapons, armor, implants, and biotic abilities. Some RPGs have found ways to endow this convention with interest; *Mass Effect* does not, and its available upgrades have a relentless similarity. Early in the game, for instance, you find an Avenger assault rifle. Later you find an Avenger II. Then an Avenger III. A different assault rifle, the Banshee, can also be found. So can the Banshee II. And the Banshee III. Additionally, weapons, armor, weapons upgrades, and armor upgrades are the only functional items you find in the world of *Mass Effect.* On one mission, you find the ancestral armor of one of your squad mates, which, for some reason, cannot be worn. This is heresy from the usual treasure-hunting practice of the RPG, with books, notes, photos, and other items scattered throughout the gameworld. The whole reason you hunt for unusual items in RPGs, after all, is to *use* them. As one critic who otherwise admired *Mass Effect* pointed out, this "detracts from the realism of the world. Imagine driving through a desolate ice field on a distant planet, picking up some debris on your scan, making your way to it and finding an old crashed probe, and finally, opening it up to find . . . a sniper rifle."

Furthermore, *Mass Effect,* like many RPGs, places a limit on the amount of gear you can carry, which is, again, convention. Figuring out what to drop and what to hold on to is one of the RPG's especial challenges. (While playing the RPG *Oblivion,* I accidentally discarded my beloved Duskfang sword while standing on the edge of a waterfall, the Niagaracal rush of which claimed it. Since I had not saved my progress for a while, I dived into the lake at the bottom of the waterfall to search for the irreplaceable Duskfang. My search lasted close to half an hour. I then bit my lip and reloaded my last saved game.) Several games have creatively approached the matter of inventory management. *Fallout 3* allows you to carry more than you are able to, but this also slows your pace to a crawl. *Resident Evil 4* forces you to arrange your gear so that it fits into a briefcase, which winds up feeling exactly as stressful as packing. *Mass Effect* limits your inventory but the limit is so absurdly high that, when it is finally reached, and you have to figure out what now to do with your thirty-seven sniper rifles, your first impulse is to turn one of them on yourself.

The best part of many RPGs is wandering the gameworld and seeing what happens. Many gameworlds are arranged in a way that allows the gamer an almost subliminal sense of where to go: a hallway bathed in reddish light will rarely provide a way out; a hallway bathed in greenish light often will. RPGs frequently neglect to provide such markers in order to encourage exploration, and the gamer often comes to have a bizarrely eidetic familiarity with gameworld landmarks. Any gamer trying to describe to another where something can be found in an RPG will often have directions as longitudinally inviolate as those of a real map: "You know that room with the two guys standing out in front of the door, beneath the staircase by the elevator? Yeah, go straight through it, around that weird partition, and you can find the colony administrator in the back, sitting at his desk." (I am frequently startled by how well I remember certain gameworlds: which crate to look in, which turn to take, which corner has an enemy around it, where to pause to reload. I often wonder where these mental maps reside in my mind. The same place where I have stored my extensive understanding of Lower Manhattan or my sketchier grasp of central Paris? I was once able to find my way from London's Trafalgar Square to the British Museum based solely on my experience of playing Team SOHO's open-world driving game *The Getaway,* so perhaps so.) Unfortunately, *Mass Effect*'s achievement in the exploration area is middling. Although the mission-critical planets the gamer must visit are all well designed, full of interest, and quite pretty, the mission-optional planets are pedestrian. There is the snow world, the sylvan world, the rock world, and the other rock world. The skies are different only in terms of their color and texture; cloud patterns are frequently identical. The mission-optional planets are also so underpopulated that they appear to have been neutron-bombed. The biggest problem, however, is that you get around these planets by driving a six-wheeled vehicle known as the Mako, which handles about as well as a luxury cruise liner. Its cannon is also about as combat-effective as a luxury cruise liner: Shooting something found on an incline even a few degrees lower than your position is frequently impossible. While driving around in the Mako, you often encounter a giant burrowing worm called the Thresher Maw. Why this huge carnivorous worm is found on so many planets, and why these planets do not seem to have any other life-forms, is not really explained, unless the Thresher Maw eats everything, in which case, mystery solved. When the Thresher Maw appears, the best but most tormentingly repetitive tactic is to drive around in a circle, shooting it with the dipsomaniacally inaccurate Mako cannon. It is very easy to dislike *Mass Effect* during such moments—but then some wonderfully odd and redeeming thing will occur, such as my chance discovery of a strange, solitary, bisonic creature on the planet Ontarom in the Newton System. The game refers to this beast as a "shifty-looking cow," but it seems harmless enough—until you put your back to the thing, whereupon your credits begin to drain. It is difficult to hold anything against a game with an alien cow pickpocket.

With the above flaws acknowledged, and only partially excused, it may be hard to understand why I spent eighty hours playing *Mass Effect.* One of the best explanations I have concerns the RPG convention of incidental encounter, which *Mass Effect* integrates into its narrative more seamlessly than any other RPG I have played. In the parts of the gameworld that are densely populated, *Mass Effect* becomes an aquarium of possibility. All around you people are talking and having conversations, and one of *Mass Effect*'s nicest touches—if one is lucky enough to have a decent home stereo system—is the way the game tracks these overheard conversations, which float from one speaker to another in startlingly realistic transit. "Overheard" conversations in novels and films invariably sit at the center of inwardly pointing arrows neon with authorial portent, but *Mass Effect*'s overheard conversations—even though they are cued to begin with some interest-snagging topic or crux sentence—really *are* overheard. RPGs that lack *Mass Effect*'s ear for dialogue are often written too broadly for any sense of potential gamer agency to take hold, in which cases *interactivity* becomes a synonym for "cudgel." In *Fallout 3,* for instance, characters you walk past will sometimes turn and look at you, as though in expectation of a greeting. You can fulfill that expectation or keep walking. In an open-world game such as *Fallout 3* the narrative must be open enough to accommodate a number of contextual possibilities (have you just been caught attempting to pick that character's pocket? have you recently injured someone he cares about?), but turning around to have a few words typically results in two kinds of encounters, one frustratingly overdetermined ("Go do this"), the other vague and desultory ("Nice day today"), neither possessing much dramatic fluidity. Free from the demands of imposed, authorial order, gameworld interactions frequently go adrift in a strange dramatic vacuum. A situation that must accommodate many possibilities may not be equipped to satisfactorily depict any of them. While playing *Mass Effect,* this criticism rarely feels applicable.

My favorite incidental encounter in *Mass Effect* involves an alien race called the hanar, also known as "jellies," which resemble the flying toasters of early-screensaver fame. The hanar believe in something called "the Enkindlers" and are persistent evangelists on their behalf. At one point a stroll along a Citadel promenade leads to your overhearing an argument between a hot-gospeling hanar and a turian police officer, who maintains, quite reasonably, that the hanar needs an evangelical permit if he wants to shout news of the Enkindlers. If you choose to intervene, your available responses range from kind (you buy the permit for the hanar), to the Hitchensian (you tell the jelly to scram and have a few rough words with the turian at his expense), to the observant (you argue with the turian in favor of the hanar's faith). This is one of several *Mass Effect* interludes that allow you to give voice to Shepard's religious beliefs, which can run the gamut. I found it difficult to have my Shepard say anything even remotely pro-religion; I took the Hitchensian position every time, despite the fact that during my various *Mass Effect* play-throughs I have experimented with just about every possibility allowable. I did not blink in the moment I allowed Shepard to procure drugs for an addict or backstab a grieving husband, yet I could not bring myself to buy the hanar a permit or make for it an ecumenical plea. Games such as *Mass Effect* allow the gamer a freedom of decision that can be evilly enlivening or nobly self-congratulating, but these games become uniquely compelling when they force you to the edge of some drawn, real-life line of intellectual or moral obligation that, to your mild astonishment, you find you cannot step across even in what is, essentially, a digital dollhouse for adults. Other mediums may depict the necessary (or foolhardy) breaches of such lines, or their foolhardy (or necessary) protection, but only games actually push you to the line's edge and make you live with the fictional consequences of your choice.

•

A late *Mass Effect* mission involves an assault on an enemy strong-hold. While you are discussing your strategy, an unexpected reve-lation of what is inside the stronghold causes one of your squad mates to object to the mission's objective, which could place the survival of his race in peril. Your attempt to reason with him along "the good of the many" lines causes him to march off and sulk on the strand of a nearby lake. When you walk over to talk some sense into your squad mate, the conversation quickly escalates and, suddenly, you find yourself in a Mexican standoff. The first time I played *Mass Effect,* I had grown immensely fond of the aggrieved character, and each time a new conversation option appeared I felt a noose of real dramatic concern tighten. Even though all I was doing was selecting lines of dialogue, the experi-ence of doing so became as gripping as a full regimental assault by geth.

Because I had not completed an earlier mission that would have allowed me a way out of this confrontation (a mission I was then unaware of), and because I had not invested my skill points in the areas of persuasion needed to convince the squad member to relent, the encounter ended with my squad mate dead at my feet and my mouth dropped open. I reloaded my last saved game and tried again. Once more, my friend and teammate took a bullet for his trouble. In my bewilderment I pressed on, and when the menu for squad selection came up, the slot for the dead character was now a darkened silhouette. It seemed as stark and inarguably final as a tombstone.

Later in the same mission, I was confronted with the loss of another squad member. This time there was no way out: Two of my teammates were trapped, and only one could be saved. I had been relying on the firepower and battle prowess of one of the trapped teammates throughout the game, and the other was

someone who had taken up with me two of three points of a growingly isosceles love triangle. Complicating matters was the fact
that the battle-hardened teammate had recently done something
unforgivable and I wanted the romantic relationship with the
other trapped character to be consummated. Thus the game took
my own self-interest and effectively vivisected it. When decision
time came, I literally put down my controller and stared at my
television screen.

This is how games get inside of you. Murder mysteries in which
you must hunt for clues. Revenge fantasies that force you to pull
the trigger. Science-fiction sagas in which you orbit the planet,
select your crew, and step off the landing vessel. And yes: love stories in which you have to choose. When games do this poorly, or
even adequately, the sensation is that of a slightly caffeinated
immediacy. You have agency, yes, but what of it? It is just a game.
But when a game does this well, you lose track of your manipulation of it, and its manipulation of you, and instead feel inserted so
deeply inside the game that your mind, and your feelings, become
as seemingly crucial to its operation as its many millions of lines
of code. It is the sensation that the game itself is as suddenly,
unknowably alive as you are. As I sat there trying to figure out
what to do, *Mass Effect*, despite its three-hundred-thousand-word
script and beautiful graphics, was no longer a verbal or visual
experience. It was a full-body experience. I felt a tremendous
sense of preemptive loss and anxiety, and even called my girlfriend, described my dilemma, and asked her for her counsel.
"You do know," she said, "that you're crazy, yes?" On the face of
things, she was right. Here I was—a straight, thirty-four-year-old
man—worrying over the consummation of my female avatar's
love affair. But she was also wrong. To say that any game that
allows such surreally intense feelings of attachment and projec-

tion is divorced from questions of human identity, choice, perception, and empathy—what is, and always will be, the proper domain of art—is to miss the point not only of such a game but art itself.

I made my choice. The game, nodding inconclusively, went on.

FALLOUT

HEADSHOTS

THE UNBEARABLE
LIGHTNESS OF GAMES

THE GRAMMAR OF FUN

LITTLEBIGPROBLEMS

BRAIDED

MASS EFFECTS

FAR CRIES

GRAND THEFTS

EIGHT

When I was a Catholic schoolboy, my friends and I spent many recesses playing a game called "Who Can Die the Best?" The rules, which I invented, were simple: One boy would announce the type of weapon he was holding—morning star, lance, M-80, Gatling gun, crossbow, pencil bomb, glaive—and then use it to kill the other boys around him. Whoever died "the best" (that is, with the most convincingly spasmodic grace) was declared winner by his executioner and allowed to pick his own weapon, whereupon a new round commenced.

One day the kind but bubonically halitosic Sister Marie wandered over to inquire why a dozen of her boys were rolling on the ground and screaming about lost limbs. When none of us answered, Sister Marie cagily turned and asked the same question of a nearby girl, who first summarized the rules of "Who Can Die the Best?" and then explained—Sister Marie's face, by now, glacially white with horror—that what we had been reacting to were grenades from Jeff Wanic's imaginary bandolier. "Who Can Die the Best?" was, from that moment on, as staunchly forbidden as meat on Friday. We kept playing, of course, but with weapons—hemlock, blowgun, freeze ray—that produced less spectacular death throes. Twenty-five years later I have no explanation for why pretending to kill and die was so much fun, but I do know that a boy alive in 2010 would find "Who Can Die the Best?" about as interesting as mime. To experience the dark gravitational pull of simulated death, that boy could play any number of violent video games in the nunless privacy of home.

This is not to suggest that the video games of my childhood were innocent. As far back as 1982, Surgeon General C. Everett Koop claimed video games were responsible for many obvious "aberrations in childhood behavior." A fair characterization of circa-1982 video-game violence would be "the collision and disappearance of two blocky abstractions," and this was disturbing only because it was occurring within a medium universally considered as intended for children. No one feared these games as tickets to instant pubescent frenzy; these games were the subject of long-term, *Manchurian Candidate*–type fears. Today's video games are often feared as objects of occult influence, particularly after the Columbine massacre, the perpetrators of which were said to be fans of a modded version of the classic shooter *Doom*. Any debate about game violence will almost inevitably become a debate about shooters. To many who oppose the video game, the video game *is* the shooter: A more assailed game genre does not exist. To fans of the shooter, the shooter *is* the video game: A tighter, less sororal game subculture does not exist.

Yet the shooter has been the messenger of many of the video game's most important breakthroughs. The very first shooter, Atari's 1980 stand-up *Battlezone,* used something called wireframe 3D, which I do not in the least understand but gave the game its distinctive appearance of see-through polygon tanks rumbling across see-through polygon battlefields, past see-through polygon hills. (Playing it felt a little like declaring war on geometry.) However primitive *Battlezone* looks now, the core of its wireframe technology is still used, and nearly every painted, beautifully dense object within a three-dimensional video-game world was, at one point, a see-through wireframe model. An even more important technological advance was marked by the 1992 appearance of the first first-person shooter, id Software's *Wolfenstein 3D,* which provided gamers with their alpha experience of three-dimensional video-game movement. id followed *Wolfenstein* with the equally influential FPSs *Doom* and *Quake.* id's co-founders, John Romero (a man of thermonuclear charisma and questionable moral probity) and John Carmack (considered by many to be one of the most brilliant programmers of all time), were undeniably gifted, but their games were curiously emblematic of the kamikaze heedlessness of the 1990s, and Romero showed an almost Clintonian talent for self-destruction. Nevertheless, the trail blazed—and shot, stabbed, and chainsawed—by id and its games was soon crowded with other developers seeking to design shooters of maximal mayhem supported by equally maximal technology. The shooter thus came to be known by many adamantine clichés (the grizzled and reluctant hero, the copious gore, the mid-game provision of some especially freaky weapon) and mimeographed narrative goals (steal the plans, secure the area, assassinate him, find more ammo, save her, blow up the bridge, find more health, trust him, suffer betrayal, huge final fight).

A second wave of FPSs suggested that the genre had unexpected riches to mine. In 1997 Rareware's *GoldenEye 007* (the greatest licensed game of all time and one of the greatest games of all time) proved beyond a doubt that the shooter was capable of being something other than an abattoir, with its endorsement of stealth tactics (whereby sneaking past enemies is as legitimate as killing them), its zoomable sniper-rifle scopes and body-part-sensitive damage mechanic (the combination of which brought into existence the immensely satisfying and instantly fatal long-

range headshot), its mission-optional objectives (which encouraged replay), and its more freely conceived gameworld (which did
away with the hallway-and-tunnel-centric design methodology of
previous shooters). The following year, Valve's *Half-Life* showed
that a shooter could go about its business with a puckish sense of
humor (its hero is a theoretical physicist) and real artfulness, as
can be seen from its riddance of cut scenes in favor of "scripted
events" (whereby the action does not stop, and the gamer remains
in control, during narrative-forwarding moments) and a gameworld composed not of disassociated "levels" but a continuous
series of locations with clear spatial relationships to one another.
Finally, in 2001, Bungie's *Halo: Combat Evolved*—in many ways the
apotheosis of the above games, with its conceptual sophistication
and feature-film-quality art design—showed that shooters could
appeal to just about anyone who played video games.

The rise of the shooter can be understood, in some ways, as an
acknowledgment of the video game's martial patrimony. The man
who, in 1958, created what is generally recognized as the first
game, *Tennis for Two,* also designed the timers used in the atom
bombs dropped on Japan. The 1962 creation of *Spacewar!* is partially creditable to military-industrial complex research-and-
development funds. The first video games may have grown out of
the apparatus of war and defense, but that apparatus was soon
using games for itself: *Battlezone* was modded by the US Army as a
Bradley armored fighting vehicle trainer; *Doom* was modded by
the US Marine Corps to attract new recruits; Valve's *Counter-Strike*
was used by the Chinese government to test the antiterrorist tactics of the People's Armed Police. Whether these games enhance
actual fighting competence is doubtful, but there is no question
that shooters train those who play them to absorb and react to
incomprehensible amounts of incoming information under great
(though simulated) duress.

Shooters are intensely violent, but their violence rarely disturbs me in the way that the violence of a game such as Rockstar's *Manhunt* disturbs me. *Manhunt* is, technically speaking, a third-person stealth game, but it is closer to an interactive snuff film. You hide in shadows, wait for someone to happen by, sneak up on him, and then use, say, a scythe to separate the unfortunate victim from his genitals. Two outstanding questions occurred to me the first (and only) time I played *Manhunt*. (1) Who thought this up? (2) Who wanted to *play* it? I endured a single hour of the game before turning it off; I spent another hour performing an exorcism on my PlayStation 2. *Manhunt* transforms the voyeuristic unease of the slasher film into something incriminatory. *Manhunt*'s scrotums are not even being mauled for the presumed benefit of flag or country, which is the detergent many shooters use to launder their carnage—quite effectively, too: The *Manhunt* series crapped out after two games, while the *Call of Duty* series is, at this writing, on its fifth vendible outing.

Many have argued that the shooter offers a sense of twisted consolation to those who will never experience war firsthand. My experience playing shooters in their online, multiplayer mode suggests that their allure is more complicated. A multiplayer round of many shooters is usually thick with active or recently active members of the military. (Their clipped, acronymic manner of speech is the giveaway.) When I was embedded with the Marine Corps in the summer of 2005, I found that nearly every young enlisted Marine I spoke to was a shooter addict, and most of the billets I visited had a GameCube or PlayStation 2. Was this at all spiritually akin to World War I–era soldiers keeping a copy of Homer or Tennyson at the ready? Did the shooter allow these Marines some small, orchestrated sanity within the chaos of war? When I asked a non-shooter-playing lieutenant about this, he reminded me that chess, too, is a war simulator.

I admire plenty of shooters, but after a night of shooter butchery I often feel agitated, as though a drill instructor has been shouting in my ear for five hours. Reflection and thought seem like distant, alien luxuries. I step outside to clear my head, but the information-sifting machine I became while playing the shooter does not always power down. Every window is a potential sniper's nest; every deserted intersection is waiting for a wounded straggler to limp across it. My stats screens in Dice's *Battlefield 2: Modern Combat* and *Call of Duty 4* tell me that I have killed many thousands of people. This information affects me about as deeply as looking over my three-pointer percentage in a basketball video game, and I sometimes wonder if shooters are not violent *enough*. The vomitous *Manhunt* actually made me contemplate, and recoil from, the messy ramifications of taking a virtual life. Most shooters do no such thing, offering a pathetic creed of salvation-by-M-16, in which you do the right—and instantly apparent—thing and bask in a heroic swell of music. On top of that, the shooter may be the least politically evolved of all the video-game genres, which is saying something. *Call of Duty 4* does not even have the courage to name its obviously Muslim enemies as Muslims, making them Russified brutes from some exotic-sounding ethnic enclave.

I do not mind being asked to kill in the shooter: Killing is part of the contract. What I do mind is not feeling anything in particular—not even numbness—after having killed in such numbers. Many shooters ask the gamer to use violence against pure, unambiguous evil: monsters, Nazis, corporate goons, aliens of Ottoman territorial ambition. Yet these shooters typically have nothing to say about evil and violence, other than that evil is evil and violence is violent. This was never the most promising thematic carbon to trace, and yet shooters keep doing so with as little self-questioning as a medieval monk copying out scripture. Shooter images of exploding heads and perforated bodies have been rotated in my mind so many times that nothing takes root. It is all simply light and color. Any shock is alleged. Every cry of pain is white noise. *Realism* has become a euphemism for how beautifully arterial blood gushes from chest wounds. Death has become a way to inject life into the gameworld. Murder is vitality. For the shooter, slaughter is its north, its south, its east, its west, and nothing—no aesthetic cataclysm—has forced the genre into any readjustment. The shooter goes on as an increasingly sophisticated imitation of a dubious original idea.

So I thought—until I played a shooter so beautiful, terrible, and monstrous that my faith was restored not only in the shooter but in the video game itself.

Some video-game developers cloak their headquarters in anonymity as a way to hold back the job-seeking hordes and add a degree of difficulty to fan-boy pilgrimages. The Paris-based developer Ubisoft is not so reticent. Its Montreal studio, found in the city's old textile district, is housed within an immense fired-brick building that, like a prison or urban high school, takes up an entire block; UBISOFT is branded upon two of its four sides.

Ubisoft Montreal has occupied this former clothing factory since 1997. When it moved in, it had only a hundred employees and required the use of part of a single floor. Today Ubisoft Montreal employs around two thousand people, the remainder of the building having long succumbed to the company's expansion. Ubisoft Montreal began modestly, with a focus on small and licensed games. When I asked why this was, I was told that, in 1997, hardly anyone who worked at Ubisoft Montreal had any idea what he or she was doing. Almost no one had any game-design experience at all. In spite (or, just as likely, because) of this, Ubisoft is today one of the most consistently innovative major developers in the world.

Its startling lobby looked as though a ski chalet and a Star Destroyer had crashed into each other and fused: black metal stairs, creaky cherrywood floors, a bank of gleamingly argent elevators, exposed wooden joists. Strategically mounted flatscreen televisions ran a silent reel of Ubisoft commercials. While I waited to be fetched by Ubisoft game designer Clint Hocking, I noted the number of attractive young women wandering about the premises and began to wonder if the company had expanded to include an escort service or modeling agency or both.

Hocking appeared before too long. Dressed in a thermal gray long-sleeved tee, cargo pants, and black canvas sneakers, his skull and jawline dark with stubble, Hocking was slender in the way that writers and musicians are sometimes slender: not out of any desire or design but rather because his days were spent being consumed rather than consuming. He led me through the warrens of Ubisoft, one magnetically sealed door after another popping open with a wave of his security card. We passed meeting rooms named for the cities with Ubisoft offices (Montreal, Hong Kong, Shanghai, Sydney, Tokyo, Sao Paolo, Brussels, San Francisco) and large, loftish spaces where the company's games were developed. As with many companies, each project gets its own large, loftish space in order to allow the creative team constant interaction. At the time of my visit, twenty projects were in different stages of development, and some of the rooms were busier than others.

In the *Prince of Persia* room, for instance, only a dozen or so people were at their desks, all of whom were working on the (already available) game's new downloadable content, the release date of which was approaching. *Prince of Persia,* a brilliant game that did not at all get its critical or commercial due, has the most hauntingly lithium ending of any video game since Team ICO's *Shadow of the Colossus* (which *Prince of Persia* in many ways resembles). In the penultimate scene of *Prince of Persia,* your love interest, Elika, with whom you have spent the game flirting and bickering, perishes in her successful effort to imprison a great evil. You then have two choices: restore her to life and release the evil or keep the evil imprisoned and turn the game off. I restored her to life. After the resurrected Elika sits up, she asks, grievingly, "Why?" You do not respond. As you carry her away, the world collapses behind you and the game ends, savagely undercutting Kurt Vonnegut's famous point that any story that concludes with lovers reunited is, even if a million invading Martians are headed toward Earth, a happy ending. (The *Prince of Persia* downloadable content being worked on during my Ubisoft visit would turn out to be a lengthy, somewhat pointless epilogue.)

We entered something called the Playtest Room—actually, a small, corridorlike space between two separate Playtest Rooms, on either side of which was a tinted one-way mirror. Here Ubisoft's developers watched and listened to the gameplay reactions of people pulled off the street. The room was fully miked, and for a few minutes we listened to two young men and one young woman discuss their moment-by-moment reactions to Epic's *Gears of War 2.* (Ubisoft occasionally canvasses outside opinion on rival games.) I asked if these people were aware that we could hear them. "It's probably in the fine print," Hocking said with a laugh. Next we walked by the Quality Testing Room—in which Ubisoft employees test games and game patches—and observed several dozen men and women playing various Ubisoft titles with dronelike industry. The final stop of the tour was a recently completed wing of classrooms. Here, Ubisoft employees between projects could listen to lectures on game-design theory and educate themselves about new technologies. This was intended to prevent layoffs. In all the economic turmoil of the last year, I was told that Ubisoft Montreal had not let a single employee go and had no plans to.

On our way to the meeting room where our interview would take place, Hocking paused in a stairwell and pointed up at the numerous exposed pipes. "A lot of Sam's moves came from here," he said. "Ideas about how to climb and hide and ambush people." "Sam" was secret agent Sam Fisher, lately of the National Security Agency and the hero of *Tom Clancy's Splinter Cell*, the first game Hocking worked on, which was released in 2002. Heavily influenced by the *Metal Gear Solid* series, *Splinter Cell* is a narratively intricate stealth game, the gameplay of which is founded upon ambush, shadow lurking, sneaking, and evasion. In the meeting room at last, I asked Hocking how he came to be involved with *Splinter Cell*.

Despite having grown up in Vancouver as a Commodore VIC-20 enthusiast, Hocking "kind of completely dropped out of the gaming world" from high school until well into his university years. In 1996, however, he abandoned his Mac for a PC and began to play "the hardcore PC games of the mid- to late 1990s": *Thief, System Shock, Deus Ex, Duke Nukem,* and, finally, *Unreal Tournament.* The last had a multiplayer-map-editor function with which Hocking became immediately fascinated. "That was really complicated," he told me. "I was building multiplayer maps and testing them with friends and figuring out how stuff works. I mean, there's no manual. There're no instructions on how to do this stuff. It's really, really hard to use—as difficult as learning to be an architect, I'm sure."

One day a friend of Hocking's sent him an e-mail about a job opening at Ubisoft Montreal. Qualifications: knowledge of the Unreal Engine Hocking had spent the last year figuring out. "I think he sent it almost as a joke," Hocking said of his friend. "I was like, 'What the hell?' I literally dragged my résumé into an e-mail and sent it in." Six weeks later he was living in Montreal and working on *Splinter Cell*. His good fortune had only begun. After a few months, the game's designer left the company and Ubisoft asked Hocking if he would take over. Then the scriptwriter left. Again, Hocking was asked to take over because, in his words, "I was one of the only Anglophones on the team and had a master's degree in creative writing." (This formal dramatic training sets Hocking apart from many game designers. When I asked which writers Hocking admired, he admitted to having a yen for "weird stuff," and named Thomas Pynchon and David Foster Wallace—may he rest in peace—as examples.) With these sudden and unforeseen promotions, Hocking was the point man for what Ubisoft hoped would become a flagship title. These hopes were fulfilled: *Splinter Cell* was, in Hocking's words, a "megahit." Recognizing Hocking's talents, Ubisoft soon asked him to serve as one of the Montreal studio's creative directors, a job he has held ever since. Of these startling turns of event, Hocking remained circumspect: "How many thousands of guys got their first job in the game industry and worked on a game that got canceled, or was a piece of shit, or no one ever played? I landed on the right game at the right time."

After Hocking completed the *Splinter Cell* sequel, *Splinter Cell: Chaos Theory,* he was asked to revitalize the Ubisoft first-person-shooter series *Far Cry* (though the developer of the original 2004 PC *Far Cry* title was the German company Crytek). The *Far Cry* series was notable for its visual beauty, paucity of load screens while moving around its South Pacific locales (even when traveling in- and outdoors, which was and remains unusual), the inspiration it siphoned from H. G. Wells's *The Island of Doctor Moreau,* and not much else. The series had been marred by its umpteen, increasingly colonic iterations: the Crytek PC game *Far Cry* was followed by the Ubisoft-developed Xbox remake, *Far Cry Instincts,* which was followed by an Xbox sequel, *Far Cry Instincts: Evolution,* which was followed by an Xbox 360 remake of the two titles bundled together, which was called *Far Cry Instincts: Predator.* Rather than stick the gaming audience with *Far Cry 5,* or *Far Cry Instincts 3: Predator 2,* Hocking pushed to call the game *Far Cry 2,* even though it had almost nothing in common with the original *Far Cry.* It was the first of many wise decisions.

Before *Far Cry 2* begins, you peruse what appear to be the case files of nine male mercenaries. The one you select will be the character you control for the game's duration. These gentlemen include a Chinese from Xinjiang, a Sikh, a Kosovar Albanian, a Native American Oklahoman, a Haitian, and a Northern Irishman. All are former smugglers, bodyguards, paramilitary insurgents, or military contractors. This unsavory roll call does not initially sparkle with originality. Then it dawns on you that all of these men have a historical connection to some kind of colonial conflict, whether distant or contemporary. And how many video games have you played that know what a Sikh, much less Xinjiang, even is?

So . . . the Haitian? Now you find yourself, with a first-person view of yourself, sitting in the backseat of a Jeep. The purview of most FPSs allows you to see, at most, the parts of your hand that come into contact with your weapon, but while seated in this Jeep you are able to look down at your chest and legs and over at the seat next to you, upon which lies a map and what appears to be a passport. You are aware of your mission (to kill an arms dealer known as the Jackal), but not much else. You do not even know where you are going. All you know is that you are in a troubled, unnamed African country. Your young driver, meanwhile, is starting the Jeep and apologizing for the delay. From him you learn that you are headed to a hotel in a nearby town called Pala.

You are soon chauffeured through countryside so topographically compelling and biologically aswarm with life that you may be forced to remind yourself: *This is a video game, not a safari.* What you see is an azure-skied afternoon—the sort of day in which the range of human visibility can conceivably compete with that of the divine. The dirt road you travel wends diligently toward the horizon. On the road, tire-squashed piles of animal dung. Along it, wire-fence guardrails anchored by old truck tires. Around it, cropless khaki waves of the breeze-blown savanna. The zigzag trunks of acacia trees are like lightning strikes from thunderheads of foliage. In the air, flitting cruciform dragonflies. In the distance, anciently knobby rock hills ringed with tonsures of greenery. Above, a plane noisily banks and grows more quietly distant—the last such plane, your driver tells you, out of this country.

This sunlit world suffers a grim and abrupt eclipse. Some Africans are walking toward you, toward the airport, seeking escape. Your driver beeps at them but sadly promises you that they will be disappointed. On your right a river comes sparklingly into view. Later you come across a stretch of savanna that, along with several acacia trees, is angrily ablaze with the most realistic fire effect you have ever seen in a game. The driver's radio is tuned to something called Liberation Radio, the deejay of which announces, "Speaking the truth for the truth seekers. Beware the evil APR scourge!" The driver flips the radio off: checkpoint ahead. "They're not fans of the deejay," he says. These are, apparently, gunmen employed by the UFLL, the APR's rival militia. Many armed black men quickly surround the Jeep, but it is a white with an Afrikaans accent who steps forward to speak. He is curious about you, but your driver douses that burning fuse by promising to bring the men cold beer on his way back. Once the Jeep is waved along, your driver showers the white man with unctuous gratitude: "Yes, sir. Thank you, sir. See you soon, sir." The moment the checkpoint is cleared, he mutters, "Foreigners." Quickly he turns to you. "No offense, sir." Moments later you see several African men standing before a row of tin-roofed shacks to which they have apparently just set fire. They stare at you ominously as you float by. The driver, waving away the smoke, says, "Don't let this concern you. Just boys letting off steam. You remember how it is."

I have traveled to a few places in which everyone was, to one degree or another, worried about being violently killed, and I have traveled to other places in which the threat of violence is always in circular, vulturine motion. I have also traveled in Africa. The driver's affected naïveté, the cable-knit menace of the checkpoint, the helixical entwinement of seeming normality with imminent collapse: All of this rings very true to me. The details scattered throughout this sequence of *Far Cry 2*—the longest scripted sequence in the game—do not tell a story, or introduce any characters, or establish any ammo dumps of plot. Because the gamer is in control of the camera, there is no establishing shot and no slow pan. Nor are there any music cues. Video games are very good at using detail to induce awe, but *Far Cry 2* understands how smaller details cytoplasmically gather around a moody nucleus of place.

You do not see your driver again. You quickly fall ill with malaria and wake up in your hotel room to find the Jackal reading aloud your assassination orders. Rather than kill you, he tells you of "a book I read a long time ago," which he proceeds to quote: "Life itself is will to power. Nothing else matters." After plunging his machete into the wall above your head, the Jackal leaves you there. A gunfight swiftly erupts outside the hotel, which you must now escape. Once this is done, you will spend many hours running errands for fatuous African revolutionaries and forging dangerous relationships with fellow mercs—the very men whose case files you initially perused and passed over for your Haitian. (Had you picked someone else, the Haitian would be among them.) These mercs—whom the game refers to as "buddies" and not, note, "friends"—will frequently request your aid with matters that *dirty work* will not begin to describe. All of them will eventually betray you and you will betray them. Others will hunt you. You will hide and run. You will kill and do other unspeakable things. And you will do your best to ruin, burn, and otherwise destroy one of the most beautiful gameworlds ever created.

Far Cry 2 is not a game about story or character. It is not a game about choice, since almost all the choices it gives you are selfish or evil ones. It is, instead, a game about chaos—which you enable, abet, and are at constant risk of being consumed by. At one point in *Far Cry 2*, I was running along the savanna when I was spotted by two militiamen. I turned and shot, and, I thought, killed them both. When I waded into the waist-deep grass to pick up their ammo, it transpired that one of the men was still alive. He proceeded to plug me with his sidearm. Frantic, and low on health, I looked around, trying to find the groaning, dying man, but the grass was too dense. I sprinted away, only to be hit by a few more of his potshots. When I had put enough distance between us, I lobbed a Molotov cocktail into the general area where the supine, dying man lay. Within seconds I could hear him screaming amid the twiggy crackle of the grass catching fire. Sitting before my television, I felt a kind of horridly unreciprocated intimacy with the man I had just burned to death. Virtually alone among shooters, *Far Cry 2* does not keep track of how many people you have killed. The game may reward your murderous actions but you never feel as though it approves of them, and it reminds you again and again that you are no better than the people you kill. In fact, you may be much worse.

Africa has not been visited by many video games. Those that have—such as the old stand-up *Jungle Hunt*—have fallen somewhat short of honoring it. Parts of the *Halo* series take place on Zanzibar, but this is a far-future, sci-fi Africa—not really Africa at all. *Resident Evil 5* uses its African setting as a master class in cultural sensitivity, such as when its muscle-sculpture white hero guns down (literally!) spear-chucking tribesmen. *Far Cry 2* escaped the accusations of racism that justifiably dogged *Resident Evil 5,* and I asked Hocking about the potentially controversial—not to mention commercially and aesthetically unusual—decision to set his game in the middle of a contemporary African civil war.

"We had a mandate from the company," he said, "which was to rejuvenate the brand. That meant getting away from the tropical island where the previous *Far Cry*s had been set." His design team kicked around various locales, but "the one that seemed to be the most powerful was the African savanna. Plain, acacia tree, sun, some herd animals in the background. That's what we wanted—that iconic, powerful feeling of natural wilderness, themes of man and nature and the darkness inside us, just like *Doctor Moreau.* As soon as you transplant that to Africa, you go from H. G. Wells to Joseph Conrad. We were making *Heart of Darkness* the video game. How bad will people be? And why? Let's not strip the race out of it. Let's go to Africa and treat it realistically and try to explore it."

Occasionally, *Far Cry 2* gives voice to paleo-liberal pro-Africa sentiment. "This is our struggle. Africa is for Africans," one militia leader tells you, and the Jackal turns out to be a bit more complicated than a mere arms dealer. When he contemptuously notes that a kid from Iowa who gets killed peacekeeping an African civil war earns more press coverage than fifty thousand dead Africans, the Jackal sounds like Noam Chomsky with a Mac-10. Such sentiment does not last long against the game's thanatological tidal wave, and it is not supposed to.

What *Far Cry 2* explores is not why civil wars occur, or why people engage in evil behavior, or why Africa is so bewilderingly prey to the wicked aims of a few. It explores, in gameplay rather than moral terms, the behavioral and emotional consequences of being exposed to relentless violence. Most shooters only play at making the gamer feel truly assailed. Even in excellent shooters, such as the *Gears of War* series, the firefights are subject to control: Here is the part where five Locusts spawn, and here is the part where two Boomers appear, and here is the part where five Wretches skitter toward you along the metal catwalk. In *Gears,* at the end of a skirmish, an electric-guitar power chord—G, I think—rings out to let you know you have cleared the sequence of enemies, are momentarily safe, and can now move on. In the *Call of Duty* games, the fighting is even more rigorously controlled. During many gunfights, you find yourself sprinting across the battlefield to the invisible spot on the game map that deactivates the spawning mechanism filling the overhead window with snipers.

In *Far Cry 2* none of the firefights is scripted. While you have missions, the gameworld is open, and you can travel—by boat, Jeep, car, bus, hang glider, or foot—virtually anywhere you want within its fifty-square-kilometer area, with carefully placed cliffs, rock formations, mountains, and rivers there to complicate things. Militias patrol, independent of where you are in the story. They are always out there, looking for you, and all of them want you dead. Checkpoints are numerous, and if you are as much as glimpsed by those manning them you will be attacked en masse. When a battle in *Far Cry 2* begins you have no idea what will happen. Simply because you are fighting five guys does not mean that four more cannot show up and join in. And simply because you successfully make it through one grueling encounter does not mean another will not happen ten seconds later. When *Far Cry 2* is set on its highest difficulty level, it becomes as thrillingly challenging as any game I have played. You learn to love and fear the violence in equal measure. Then you listen to a man burning to death in the grass and the game lets you make of that what you will.

I described to Hocking my most memorable *Far Cry 2* experience. It was morning—the game has an evocative day–night cycle, with morning suns as bright as magma and night skies buckshot with stars—and I was driving along in a Jeep, on my way to steal a bag of the African narcotic *khat* for one of my fellow mercs. Around the bend ahead of me came another Jeep. Since almost everyone in the world of *Far Cry 2* wishes you ill, I began to think about my course of action. Suddenly a zebra—perhaps headed for the local watering hole for its morning refreshment—ran into the road. I swerved, unsuccessfully, to avoid it and smashed grille-to-grille into the other Jeep, the passenger of which had opened up on me with a .50-caliber mounted machine gun. The zebra, meanwhile, was pinned between us. I jumped out of my vehicle and spun and lobbed a grenade under the Jeeps and will not soon forget the surreal clarity with which the luckless zebra's blast-launched corpse went sailing past me. It was as bizarre as anything I had ever seen in a video game—and no one had written or programmed one moment of it. I asked Hocking if, when he played *Far Cry 2*, he ever felt as if he were at the Frankensteinian mercy of some incomprehensible beast of his own creation.

"I have that feeling all the time," he told me. "The sort of sublime violence and chaos—something rises up out of it that's shocking. The *beautifulness* of its horror. It's incredible how volatile and intense this game can be." With *Far Cry 2*, Hocking said, unrelentingness was "a big part of our goal. Playing the game, I've *learned* things about myself. Trying to hold your ground against fourteen or fifteen guys when you're hidden behind a Jeep with an assault rifle? Your brain is telling you, *If you get up and try to run, you're dead.* You're trying to stay calm: *I don't have enough ammo to miss, so I have to aim at these fucking guys and make sure they're down.* In most games you just kind of charge forward. There's no real tactics, no real discipline under fire. We wanted to make sure that the potential for that insane stuff was so high that you don't need to script anything because there would *always* be insane stuff happening."

He sat up in his chair, then, roused by the memory of the first time *Far Cry 2* was publicly shown at a game convention in Leipzig, Germany. "Alain Corre, the head of Ubisoft marketing in Europe, came to see the game; he hadn't seen it yet. It's a very hard game to demo, because nothing is scripted, so you're always improvising. We were showing him the game and at one point the guy who was playing got into a bad spot. We're like, 'Ah, fuck. He's gonna die in the demo. Great.' So he's just spraying machine-gun fire and hits this propane tank that was in a weird fluky position. It went flying through the air and the jet caught the three guys who were shooting at him on fire—all in a row. They *burst* into flames. The propane tank goes sailing out of frame. Then another one caught fire and goes bouncing across the ground and up into a tree, and right when it hit the branches of the tree, the tank exploded. And the tree goes, *BOOOF!* I've never seen a tree come apart in the game like that since. Every branch smashed off the tree in this huge puff of leaves. Because that tank had bounced across the ground and up into the tree, there was this huge line of scorch marks where it had spun through the grass. It was so amazing, this crazy chain of events—and to this day I've never seen a chain reaction that was so cool. Everything that happened was totally systemic. There was no way we could have scripted that. And then Alain turned to us and said, 'Your game is great.' "

Hocking is obsessed with the formal reverberations of game design. "I *despise* cut scenes," he told me. "We have a mandate, actually, not to use cut scenes. It's not necessarily engraved in stone, but most of us believe we need to try to tell a story in an interactive way." When I brought up Rockstar's *Grand Theft Auto IV*, the cut scenes of which are generally superb, Hocking nodded and said, "As a player I don't necessarily dislike them if they're done very, very well. As a developer, on the other hand, the cost of them is so high. The constraints that they bring are significant. Once a cut scene is built and in the game, you can't change it. You're done. A lot of my work on the original *Splinter Cell* was building cut scenes, which is a massive waste of time. They were taking my time away from making the game more fun."

Far Cry 2's maintained first-person point of view tries to supplant the need for cut scenes. Every in-game activity—talking to someone, studying your map, climbing into a car, opening a door, using pliers to remove a bullet from your leg, relocating a broken thumb, popping an antimalaria pill—holds the first-person and places no freeze on the surrounding action. (Hocking swiped this from DreamWorks Interactive's *Jurassic Park: Trespasser*, an ambitious but largely unsuccessful game.) *Far Cry 2*'s devotion to its unbroken first-person point of view may not sound unusual or even noteworthy—until you find yourself running from seven militiamen and trying to consult your map while suffering a simultaneous attack of vision-blurring malarial fever.

The maintained first-person was intended to provide what Hocking calls a psychosomatic "shortcut" to the gamer's brain.

"The reason is twofold. First, you create this bond between the player and the character. When he has to pull a twig out of his arm, he feels some kind of illusion of pain. Second, all of it was designed to build up to that moment when you're holding your buddy in your arms. It's this huge chain of connectedness that pays off in that moment."

Hocking was referring to moments in *Far Cry 2* in which one of your merc buddies attempts to come to your rescue, only to be cut down on the battlefield. When you approach severely injured buddies, they beg you to help and curse you if you put your back to them. If you choose to help, you take your wounded buddy in your arms. You then have three choices: abandonment, healing, or mercy killing. If you pull out your sidearm but hesitate to fire, your buddy will sometimes grab the gun barrel, place it to his lips, and demand that you put him out of his misery. These moments are unnerving not because your buddies are deeply imagined characters. They are types, nothing more. What *Far Cry 2* seeks to provide with depth is the actual, in-game experience of terminating a life or being the agent of its restoration. This is not a tragic choice, and does not pretend to be. It is a way to lure you deeper into the gameworld's brutal ethos.

That you can hurt your buddies at all runs counter to the way most shooters deal with friendly characters, who are either magically immune to your bullets or whose death by your hand results in instant mission failure. On this point Hocking grew animated: "I guarantee that the first time you went into one of those interactive scenes in *Half-Life,* where you had your gun in your hand and were able to point it at Alyx, the first thing you did was line her up and shoot her in the head. And it *didn't do anything.* It fired, and you lost your ammo, but the bullet wasn't there. And then for the rest of the game you never questioned it. All we did was say, 'Okay, what if you can shoot that person? What happens if this dude shows up to save your life and you turn around and pop him in the forehead? What does that mean?' "

With some sorrow I admitted to Hocking that I did not register my buddies' passing quite as viscerally as he intended—and I shot a lot of them. Some final emotional tumbler refused to fall into place. Whether this was because the buddies are horrible people, or because *Far Cry 2* forgoes the clumsy (but possibly necessary) means of characterization found in other games, I was not sure. Hocking admitted the moment "didn't work as well as we hoped," and attributed it to, among other things, how "wooden" the buddies are during interaction. This would mean that my failure to be moved had more to do with the current capability of the video game than *Far Cry 2*'s failure to realize its vision. Possibly it was both.

Just as possibly, Hocking said, it was neither. "I always assume minor technical problem X prevented us from achieving perfection. I always think, *If we just solved X, we would have succeeded in everything beyond X*. But you don't actually know what is behind X. The wall behind X might be impassable." He went on to agree with the most common complaint lodged against *Far Cry 2*: Only rarely do you have the faintest fucking clue as to what, narratively, is going on. You have your buddies, but they are as fickle and unforthcoming as house cats, and your interaction with them is fleeting. On top of that, you are working for two different militia factions simultaneously, which neither appears to mind. (They also prepay you for your work, which resembles no merc protocol of which I am aware.) For long stretches of the game you do not know who is related to whom or how or why anyone happens to be doing what they are doing.

"That stuff," Hocking said, "is being tracked, but it's all just a bunch of invisible matrices that aren't exposed to the player. It would have been in our interest to make the game more about the aggregate relationships of all the people and expose who likes who and who dislikes who—making clear to the player that all of these people are interconnected." A less formally adventurous game would have provided such information in typical video-game fashion: either artificially, such as a chart with the faces of the various characters beside mood-indicating glyphs, or with the marginal dynamism of allowing the gamer to find some miraculously thorough memo. Hocking could not bring himself to resort to such conventions.

I asked him, "So how *do* you reveal that information?"

"I don't know," he said. "That's the question. And the problem."

Like approximately everyone else on Planet Earth, Hocking writes a blog. His most famous and impassioned post, "Ludonarrative Dissonance in *BioShock*," appeared in late 2007. Although Hocking opens the post by assuring his readers that *BioShock* is an indisputably great game, he says that, as a game designer, he is unable to overlook its central failure. *BioShock* invites "us to ask important and compelling questions," Hocking writes, but the answers it provides "are confused, frustrating, deceptive and unsatisfactory."

Among the games of this era, *BioShock* has Himalayan stature. From writing to level design to art direction to gameplay, it is a work of anomalous and distinctive excellence. The story takes place in 1960, in an underwater city known as Rapture, secretly designed and just as secretly overseen by a wealthy madman named Andrew Ryan. Ryan is a cleric of a philosophy clearly intended to resemble the Objectivism of Ayn Rand, and he runs his city accordingly. Rapture is a place, Ryan says, where science is not limited by "petty morality," where "the great are not constrained by the small." A mid-Atlantic plane crash leads to the gamer's unplanned, many-fathomed descent into Rapture, which has been recently torn apart by riots and rebellion; its few surviving citizens are psychopathic. Thanks to a two-way radio, a man named Atlas becomes your only friend and guide through the ruins of Ryan's utopia, and your first task is to help Atlas's family escape.

The only thing of any worth in Rapture is ADAM, an injected tonic that grants superhuman abilities such as pyro- and telekinesis. One way to gorge on ADAM is by "harvesting" small, pigtailed girls—known as Little Sisters—who wander Rapture under heavy guard. But Little Sisters can also be restored to uncorrupted girlhood. Thus, whenever the gamer comes into contact with a Little Sister, he must decide what to do with her. This involves taking the girl into your arms. If the gamer saves her, the reward is a moon-eyed curtsy of thanks. If the gamer harvests her, the Little Sister vividly and upsettingly transforms into a wiggling black slug—but the reward is ADAM, which makes you more powerful.

"*BioShock,*" Hocking writes, "is a game about the relationship between freedom and power. . . . It says, rather explicitly, that the notion that rational self-interest is moral or good is a trap, and that the 'power' we derive from complete and unchecked freedom necessarily corrupts, and ultimately destroys us." The problem is that this theme lies athwart of the game itself. For one thing, there is no real benefit in harvesting Little Sisters, because refusing to harvest them eventually leads to gifts and bonuses of comparable worth. In other words, the gamer winds up in a place of equivalent advantage no matter what decision he or she makes. *BioShock* was celebrated for being one of the first games to approach morality without lapsing into predictable binaries, but if the altruistic refusal to harvest Little Sisters has no sacrificial consequence, the refusal cannot really be considered altruistic.

For Hocking, this was only the beginning of *BioShock*'s dissonance: "Harvesting [Little Sisters] in pursuit of my own self-interest seems not only the best choice mechanically, but also the right choice. This is exactly what this game needed to do—make me experience—feel—what it means to embrace a social philosophy that I would not under normal circumstances consider." But *BioShock,* Hocking argues, does not follow through with this, as it is designed in such a way as to force the gamer to help Atlas. This does not make sense "if I am opposed to the principle of helping someone else. In order to go forward in the game, I must do as Atlas says, because the game does not offer me the freedom to choose sides." While the game's mechanic offers the freedom to luxuriate in Objectivism's enlightened selfishness, the game's fiction denies the gamer that same freedom.

Because helping Atlas is "not a ludic constraint" but rather "a narrative one that is dictated to us," Hocking claims to have felt mocked by *BioShock,* as though his contract of belief in the game-world were torn up before his very eyes. The post concludes with Hocking's acknowledgment that his concerns "may seem trivial or bizarre" and that he "only partially" understands his reaction to *BioShock.* "It is," he writes, "the complaint of a semi-literate, half-blind Neanderthal, trying to comprehend the sandblasted hieroglyphic poetry of a one-armed Egyptian mason."

When I asked Hocking about "Ludonarrative Dissonance in *BioShock,*" he said he did not have much to add. He admired *BioShock* and was ambivalent about *BioShock.* In the end, he said, "I excuse the fact that I don't have agency in the story for real, because I know it's a game and I know it is technologically impossible for me to have the kind of agency this game wants me to feel like I have. In a game today, it doesn't exist."

Perhaps, I said, that was the point? Rather than mock the gamer, *BioShock* could just as easily be commenting on itself, its game-ness, thereby allowing the gamer to feel what he or she wants to feel. When I played *BioShock,* I felt better *emotionally* rescuing the Little Sisters and would not have stopped doing so even if I was aware that my sacrifice was not a real one. When *BioShock* told me that I was, in fictional terms, being controlled, I thought hard about the last three days I had spent manipulating photons with a button-encrusted plastic brick. Was not the point that *BioShock* is rich enough to provoke such divergent interpretations? And was not *Far Cry 2* guilty of a ludonarrative dissonance of omission by neglecting to populate its gameworld with civilians and other innocent people? If the theme of *Far Cry 2* is the seductive and perverting power of learning how to navigate and prosper within a violent world, why is the gamer denied situations that would truly test the limits of that seduction?

Hocking immediately said, "I don't know. We didn't want to be muddying up our themes with a bunch of mass murder for laughs. That would have made it confusing. But more than that, there are social-responsibility issues involved in being able to butcher women and children." I had read an interview with Hocking in which he had described the "numerous technical and production challenges" that eliminated *Far Cry 2*'s intended civilian and refugee populations, so I knew it was not only an issue of social responsibility. Here, then, was one of the great vexations of the video game: For games to mean something, they must engage with meaningful subject matter. The subject matter need not be death and slaughter, but if it is, you must ask yourself, as a game designer, how far you are willing to let the gamer go, and why. As technology improves, Hocking said, video-game characters "are going to be so real and believable that when you shoot them with a .556 round, their arms are going to pop off. And it's going to be horrifying. No one wants that. I don't want to play that game." The place where game interactivity and visual fidelity intersect is a kind of moral crossroad at which any sane person would feel obliged to pause.

And here was the other problem. In video games, the assignation of meaning has traditionally seesawed between the game's author (or authors) and the gamer. Authors had their say in static moments such as cut scenes, and gamers had their say during play. There is no doubt that this method of game design has produced many fine and fun games but very few experiences that have emotionally startled anyone. For designers who want to change and startle gamers, they as authors must relinquish the impulse not only to declare meaning but also to suggest meaning. They have to think of themselves as shopkeepers of many possible meanings—some of which may be sick, nihilistic, and disturbing. Game designers will always have control over certain pivot points—they own the store, determine its hours, and stock its shelves—but once the gamer is inside, the designer cannot tell the gamer what meaning to pursue or purchase. The reason this happens so rarely in games—and struggles to happen even in games that attempt to follow this model—is because, as Hocking said, "The very nature of drama, as we understand it, is authored. Period. The problem is, once you give control of that to a player, authorial control gets broken. Things like pacing and flow and rhythm—all these things that are important to maintaining the emotional impact of narrative—go all willy-nilly. The player's vision of what might be dramatic or interesting or compelling can be completely at odds with the author's vision. The whole point of a game is that players have autonomy to do what they want. It might be that the player is motivated to do the opposite of what you want him to do. That's a legitimate goal in gameplay."

Hocking was perfectly aware that the social responsibility he felt in preventing the gamer from massacring civilians in *Far Cry 2* was a contradictory riptide within his overall design philosophy.

He brought up Jonathan Blow's core criticism—games that do not attempt to harmonize meaning with gameplay cannot succeed as works of art—and said, "I think he's right in the sense that a lot of people don't understand dynamical meaning—meaning that arises out of mechanics—because no one really understands it. *I* don't even know what that means and it's my medium of expression! Most people don't understand it because they can't understand it, not because they haven't taken the time to learn it. They're grappling with it like everyone else. Some people grapple with it more seriously and recognize that it's a central and fundamental flaw in what we do. I think Jon and I see eye-to-eye on that. Finding a way to make the mechanics of play our expression as creators and as artists—to me that's the only question that matters."

Where, I asked, did that leave the narrative game? And *could* narrative games ever reach a place that was artistically satisfying for their creators and emotionally significant for their players?

Hocking's hand traffic-copped. "Whoa. I'm not committed to the idea of the authored narrative game. In fact, I'm totally against it. I'm committed to the idea of designing a system wherein you provide useful channels for the player to poke and prod, so that you're kind of baiting him into narrative paths of his own choosing." *Call of Duty 4*, he said, was an example of "a rigidly authored narrative game that has a fairly good story. It pushes some of your buttons and manipulates you and makes you feel stuff. And yet the story you experienced is exactly the same as the one I experienced, with very minor variations that are probably no more different from the minor variations you and I have in our subjective experience of reading a novel. The problem with that approach, in my opinion, is that we already know how good that can be. The best story *Call of Duty* can ever have is something either very close to or marginally better than the best war movie ever made. The best we can ever hope for, with a narrative game, is to get there. We can't go beyond it using the tools of film or literature or any other authored narrative approach. The question is, can we go beyond it, way beyond it, to completely different realms, by using tools that are inherent to games? To let the player play the story, tell his own story, and have that story be deep and meaningful? We don't know the limit to that problem. It could be that the limit to that problem is stories that aren't nearly as good."

"But you've got to find out," I said.

"Yeah. I have to find out."

FALLOUT

HEADSHOTS

THE UNBEARABLE
LIGHTNESS OF GAMES

THE GRAMMAR OF FUN

LITTLEBIGPROBLEMS

BRAIDED

MASS EFFECTS

FAR CRIES

GRAND THEFTS

NINE

O nce upon a time, I wrote in the morning, jogged in the late afternoon, and spent most of my evenings reading. Once upon a time, I wrote off as unproductive those days in which I had managed to put down "only" a thousand words. Once upon a time, I played video games almost exclusively with friends. Once upon a time, I did occasionally binge on games, but these binges rarely had less than a fortnight between them. Once upon a time, I was, more or less, content.

"Once upon a time" refers to relatively recent years (2001–2006) during which I wrote several books and published more than fifty pieces of magazine journalism and criticism—a total output of, give or take, forty-five hundred manuscript pages. I rarely felt very disciplined during this half decade, though I realize this admission invites accusations of disingenuousness or, failing that, a savage and justified beating. Obviously, I *was* disciplined. These days, however, I am lucky if I finish reading one book every fortnight. These days, I have read from start to finish exactly two works of fiction—excepting those I was not also reviewing—in the last year. These days, I play video games in the morning, play video games in the afternoon, and spend my evenings playing video games. These days, I still manage to write, but the times I am able to do so for more than three sustained hours have the temporal periodicity of comets with near-Earth trajectories.

For a while I hoped that my inability to concentrate on writing and reading was the result of a charred and overworked thalamus. I knew the pace I was on was not sustainable and figured my discipline was treating itself to a *Rumspringa*. I waited patiently for it to stroll back onto the farm, apologetic but invigorated. When this did not happen, I wondered if my intensified attraction to games, and my desensitized attraction to literature, was a reasonable response to how formally compelling games had quite suddenly become. Three years into my predicament, my discipline remains AWOL. Games, meanwhile, are even more formally compelling.

It has not helped that during the last three years I have, for what seemed like compelling reasons at the time, frequently up-ended my life, moving from New York City to Rome to Las Vegas to Tallinn, Estonia, and back, finally, to the United States. With every move, I resolved to leave behind my video-game consoles, counting on new surroundings, unfamiliar people, and different cultures to enable a rediscovery of the joy I once took in my work. Shortly after arriving in Rome, Las Vegas, and Tallinn, however, the lines of gameless resolve I had chalked across my mind were wiped clean. In Rome this took two months; in Vegas, two weeks; in Tallinn, two days. Thus I enjoy the spendthrift distinction of having purchased four Xbox 360 consoles in three years, having abandoned the first to the care of a friend in Brooklyn, left another floating around Europe with parties unknown, and stranded another with a pal in Tallinn (to the irritation of his girlfriend). The last Xbox 360 I bought has plenty of companions: a Game-Cube, a PlayStation 2, and a PlayStation 3.

Writing and reading allow one consciousness to find and take shelter in another. When the mind of the reader and writer perfectly and inimitably connect, objects, events, and emotions become doubly vivid—realer, somehow, than real things. I have spent most of my life seeking out these connections and attempting to create my own. Today, however, the pleasures of literary connection seem leftover and familiar. Today, the most consistently pleasurable pursuit in my life is playing video games. Unfortunately, the least useful and financially solvent pursuit in my life is also playing video games. For instance, I woke up this morning at 8 a.m. fully intending to write this chapter. Instead, I played *Left 4 Dead* until 5 p.m. The rest of the day went up in a blaze of intermittent catnaps. It is now 10 p.m. and I have only started to work. I know how I will spend the late, frayed moments before I go to sleep tonight, because they are how I spent last night, and the night before that: walking the perimeter of my empty bed and carpet-bombing the equally empty bedroom with promises that tomorrow will not be squandered. I will fall asleep in a futureless, strangely peaceful panic, not really knowing what I will do the next morning and having no firm memory of who, or what, I once was.

The first video game I can recall having to force myself to stop playing was Rockstar's *Grand Theft Auto: Vice City*, which was released in 2002 (though I did not play it until the following year). I managed to miss *Vice City*'s storied predecessor, *Grand Theft Auto III*, so I had only oblique notions of what I was getting into. A friend had lobbied me to buy *Vice City*, so I knew its basic premise: You are a cold-blooded jailbird looking to ascend the bloody social ladder of the fictional Vice City's criminal under- and overworld. (I also knew that *Vice City*'s violent subject matter was said to have inspired crime sprees by a few of the game's least stable fans. Other such sprees would horribly follow. Seven years later, Rockstar has spent more time in court than a playground-abutting pesticide manufactory.) I might have taken better note of the fact that my friend, when speaking of *Vice City,* admitted he had not slept more than four hours a night since purchasing it and had the ocular spasms and fuse-blown motor reflexes to prove it. Just what, I wanted to know, was so specifically compelling about *Vice City?* "Just get it and play it," he answered. "You can do anything you want in the game. Anything."

Before I played *Vice City,* the open-world games with which I was familiar had predictable restrictions. Ninety percent of most open gameworlds' characters and objects were interactively off limits, and most game maps simply stopped. When, like a digital Columbus, you attempted to journey beyond the edge of these flat earths, onscreen text popped up: YOU CAN'T GO THAT WAY! There were a few exceptions to this, such as the (still) impressively open-ended gameworld of Nintendo's *Legend of Zelda: Ocarina of Time,* which was released in 1998. As great as *Ocarina* was, however, it appealed to the most hairlessly innocent parts of my imagination. Ingenious, fun, and beautiful, *Ocarina* provided all I then expected from video games. (Its mini-game of rounding up a brood of fugitive chickens remains my all-time favorite.) Yet the biggest game of its time was still, for me, somehow too small. As a navigated experience, the currents that bore you along were suspiciously obliging. Whatever I did, and wherever I moved, I never felt as though I had *escaped* the game. When the game stopped, so did the world.

The world of *Grand Theft Auto: Vice City* was also a fantasy—a filthy, brutal, hilarious, *contemporary* fantasy. My friend's promise that you could do anything you wanted in Vice City proved to be an exaggeration, but not by very much. You control a young man named Tommy, who has been recently released from prison. He arrives in Vice City—an oceanside metropolis obviously modeled on the Miami of 1986 or so—only to be double-crossed during a coke deal. A few minutes into the game, you watch a cut scene in which Tommy and his lawyer (an anti-Semitic parody of an anti-Semitic parody) decide that revenge must be taken and the coke recovered. Once the cut scene ends, you step outside your lawyer's office. A car is waiting for you. You climb in and begin your drive to the mission destination (a clothing store) clearly marked on your map. The first thing you notice is that your car's radio can be tuned to a number of different radio stations. What is playing on these stations is not a loop of caffeinatedly upbeat MIDI video-game songs or some bombastic score written for the game but Michael Jackson, Hall and Oates, Cutting Crew, and Luther Vandross. While you are wondering at this, you hop a curb, run over some pedestrians, and slam into a parked car, all of which a nearby police officer sees. He promptly gives chase. And for the first time you are off, speeding through Vice City's various neighborhoods. You are still getting accustomed to the driving controls and come into frequent contact with jaywalkers, oncoming traffic, streetlights, fire hydrants. Soon your pummeled car (you shed your driver's-side door two blocks ago) is smoking. The police, meanwhile, are still in pursuit. You dump the dying car and start to run. How do you get another car? As it happens, a sleek little sporty number called the Stinger is idling beneath a stoplight right in front of you. This game *is* called *Grand Theft Auto,* is it not? You approach the car, hit the assigned button, and watch Tommy rip the owner from the vehicle, throw him to the street, and drive off. Wait—look there! A *motorcycle.* Can you drive motorcycles, too? After another brutal vehicular jacking, you fly off an angled ramp in cinematic slow-motion while ELO's "Four Little Diamonds" strains the limits of your television's half-dollar-sized speakers.

You have now lost the cops and swing around to head back to your mission, the purpose of which you have forgotten. It gradually dawns on you that this mission is waiting for *you* to reach *it.* You do not have to go if you do not want to. Feeling liberated, you drive around Vice City as day gives way to night. When you finally hop off the bike, the citizens of Vice City mumble and yell insults. You approach a man in a construction worker's outfit. He stops, looks at you, and waits. The game does not give you any way to interact with this man other than through physical violence, so you take a swing. The fight ends with you stomping the last remaining vitality from the hapless construction worker's blood-squirting body. When you finally decide to return to the mission point, the rhythm of the game is established. Exploration, mission, cut scene, driving, mayhem, success, exploration, mission, cut scene, driving, mayhem, success. Never has a game felt so open. Never has a game felt so generationally relevant. Never has a game felt so awesomely gratuitous. Never has a game felt so narcotic. When you stopped playing *Vice City,* its leash-snapped world somehow seemed to go on without you.

Vice City's sequel, *Grand Theft Auto: San Andreas,* was several magnitudes larger—so large, in fact, I never finished the game. *San Andreas* gave gamers not one city to explore but three, all of them set in the hip-hop demimonde of California in the early 1990s (though one of the cities is a Vegas clone). It also added dozens of diversions, the most needless of which was the ability of your controlled character, a young man named C.J., to get fat from eating health-restoring pizza and burgers—fat that could be burned off only by hauling C.J.'s porky ass down to the gym to ride a stationary bike and lift weights. This resulted in a lot of soul-scouring questions as to why (a) it even mattered to me that C.J. was fat and why (b) C.J. was getting more physical exercise than I was. Because I could not answer either question satisfactorily, I stopped playing.

Grand Theft Auto IV was announced in early 2007, two years after the launch of the Xbox 360 and one year after the launch of the PlayStation 3, the "next-generation" platforms that have since pushed gaming into the cultural mainstream. When the first next-gen titles began to appear, it was clear that the previous *Grand Theft Auto* titles—much like Hideo Kojima's similarly brilliant and similarly frustrated *Metal Gear Solid* titles—were games of next-gen vision and ambition without next-gen hardware to support them. The early word was that *GTA IV* would scale back the excesses of *San Andreas* and provide a rounder, more succinctly inhabited game experience. I was living in Las Vegas when *GTA IV* (after a heartbreaking six-month delay) was finally released.

In Vegas I had made a friend who shared my sacramental devotion to marijuana, my dilated obsession with gaming, and my ballistic impatience to play *GTA IV.* When I was walking home from my neighborhood game store with my reserved copy of *GTA IV* in hand, I called my friend to tell him. He let me know that, to celebrate the occasion, he was bringing over some "extra sweetener." My friend's taste in recreational drug abuse vastly exceeded my own, and this extra sweetener turned out to be an alarming quantity of cocaine, a substance with which I had one prior and unexpectedly amiable experience, though I had not seen a frangible white nugget of the stuff since.

While the *GTA IV* load screen appeared on my television, my friend chopped up a dozen lines, reminded me of basic snorting protocol, and handed me the straw. I hesitated before taking the tiny hollow scepter, but not for too long. Know this: I was not someone whose life had been marked by the meticulous collection of bad habits. I chewed tobacco, regularly drank about ten Diet

Cokes a day, and liked marijuana. Beyond that, my greatest vice was probably reading poetry for pleasure. The coke sailed up my nasal passage, leaving behind the delicious smell of a hot leather car seat on the way back from the beach. My previous coke experience had made feeling good an emergency, but this was something else, softer, and almost *relaxing*. This coke, my friend told me, had not been "stepped on" with any amphetamine, and I pretended to know what that meant. I felt as intensely focused as a diamond-cutting laser; *Grand Theft Auto IV* was ready to go. My friend and I played it for the next thirty hours straight.

Many children who want to believe their tastes are adult will bravely try coffee, find it to be undeniably awful, but recognize something that could one day, conceivably, be enjoyed. Once our tastes as adults are fully developed, it is easy to forget the effort that went into them. Adult taste can be demanding work—so hard, in fact, that some of us, when we become adults, selectively take up a few childish things, as though in defeated acknowledgment that adult taste, and its many bewilderments, is frequently more trouble than it is worth. Few games have more to tell us about this adult retreat into childishness than the *Grand Theft Auto* series.

In *GTA IV* you are Niko Bellic, a young immigrant with an ambiguous past. We know he is probably a Serb. We know he fought in the Balkans war. We know he was party to a war-crime atrocity and victim of a double-cross that led to the slaughter of all but three members of his paramilitary unit. We know he has taken life outside of war, and it is strongly suggested that he once dabbled in human trafficking. "I did some dumb things and got involved with some idiots," Niko says, early in the game, to his friend Hassan. "We all do dumb things," Hassan replies. "That's what makes us human." The camera closes on Niko as he thinks about this and, for a moment, his face becomes as quietly expressive as that of a living actor. "Could be," he says.

Niko has come to Liberty City (the *GTA* world's run at New York City) at the invitation of his prevaricating cousin, Roman. He wants to start over, leave behind the death and madness of his troubled past, and bathe in the comfort and safety of America. Niko's plan does not go well. Soon enough, he is working as a thief and killer. Just as *Lolita,* as Nabokov piquantly notes in his afterword, was variously read as "old Europe debauching young America" or "young America debauching old Europe," *GTA IV* leaves itself interpretively open as to whether Niko is corrupted by America or whether he and his ilk (many of the most vicious characters whose paths Niko crosses are immigrants) are themselves bacterial agents of corruption. The earlier *GTA* games were less thematically ambitious. Tommy from *Vice City* is a cackling psychopath, and C.J. from *San Andreas* merely rides the acquisitionist philosophy of hip-hop culture to terminal amorality. They are not characters you root for or even want, in moral terms, to succeed. You want them to succeed only in gameplay terms. The better they do, the more of the gameworld you see. The stories in *Vice City* and *San Andreas* are pastiches of tired filmic genres: crime capers, ghetto dramas, police procedurals. The driving force of both games is the gamer's curiosity: *What happens next? What is over here? What if I do this?* They are, in this way, childlike and often very silly games, especially *San Andreas,* which lets you cover your body with ridiculous tattoos and even fly a jetpack. While the gameworlds and subject matter are adult—and under no circumstances should children be allowed near either game—the joy of the gameplay is allowing the vestiges of a repressed, tantrum-throwing, childlike self to run amok. Most games are about attacking a childlike world with an adult mind. The *GTA* games are the opposite, and one of the most maliciously entertaining mini-games in *Vice City* and *San Andreas* is a mayhem mode in which the only goal is to fuck up as much of the gameworld as possible in an allotted period of time.

There is no such mode in *GTA IV* (though the gamer is free to fuck up the gameworld on his or her own clock), one suspects because the game seeks to provide Niko with a pathos it absolutely denies *Vice City's* Tommy and mostly denies *San Andreas's* C.J. All the *GTA* games have been subject to mis- and overinterpretation, and *GTA IV* is no exception. *GTA IV's* most frequent misinterpretation is that it boasts a story many credulous game reviewers deemed "Oscar-worthy," which, they said, lent Niko's plight real relevance. This is, in a word, preposterous. At its best, the story of *GTA IV* is pretty good for a video game, which is to say, conventional and fairly predictable. At its equally representative worst, *GTA IV's* story does not make much sense, unless one believes that Niko would instantly forgive his cousin, Roman, for luring him to America under boldly false pretenses, that Niko could find a girlfriend after one day in America, that people who barely know Niko would unquestioningly entrust him with their lives and drug money, and that Niko's mother would write him e-mails in English. The one narrative task *GTA IV* handles extremely well is dialogue, particularly whenever the incomprehensible Jamaican pothead Little Jacob ambles into the proceedings. A hilariously hairsplitting argument between a pair of Irish American criminals about whether the plastic explosive they are about to use to blow open a bank door is called "C4" or "PE4" feels scissored out of a Tarantino script. *GTA IV's* dialogue has no bearing on its gameplay, of course, but does make it one of very few games in which listening to people talk is not only enjoyable but sociologically revelatory.

Niko's real pathos derives not from the gimcrack story but how he looks and moves. *Vice City* and *San Andreas* were graphically astounding by the standards of their time, but their character models were woeful—even by the standards of their time. The most vivid thing about *Vice City*'s Tommy is the teal Hawaiian shirt he wears at the game's open, and *San Andreas*'s C.J. is so awkwardly rendered he looks like the King of the Reindeer People. Niko, though, is just about perfect. Dressed in striped black track pants and a dirty windbreaker, Niko looks like the kind of guy one might see staring longingly at the entrance of a strip club in Zagreb, too poor to get in and too self-conscious to try to. When, early in the game, a foulmouthed minor Russian mafioso named Vlad dismisses Niko as a "yokel," he is not wrong. Niko is a yokel, pathetically so. One of the first things you have to do as Niko is buy new clothes in a Broker (read: Brooklyn) neighborhood called Hove Beach (read: Sheepshead Bay). The clothing store in question is Russian-owned, its wares fascinatingly ugly. And yet you know, somehow, that Niko, with his slightly less awful new clothes, feels as though he is moving up in the world. The fact that he is only makes him more heartrending. The times I identified most with Niko were not during the game's frequent cut scenes, which drop bombs of "meaning" and "narrative importance" with nuclear delicacy, but rather when I watched him move through the world of Liberty City and projected onto him my own guesses as to what he was thinking and feeling.

Liberty City, many game reviewers argued, is the real central character of *GTA IV*, and here they were not wrong. The worlds of *Vice City* and *San Andreas*, however mind-blowing at the time, were also geometrically dead and Tinkertoy-ish, with skyscrapers and buildings that looked like upturned Kleenex boxes. Their size and variegation were impressive, but surfaces held little texture and lighting and particle effects were distractingly subpar. Virtually every visual element of *GTA IV*'s Liberty City is gorgeously realized, and I have never felt more forcefully transported into a gameworld than while running across Liberty City's Middle Park in orange-sherbet dusk, taking a right turn onto the Algonquin Bridge and seeing the jeweled ocean glisten in the hard light of high afternoon, or stepping out of a Hove Beach tenement into damp phantasms of morning fog. The physics that previously governed the world of *GTA* were brilliantly augmented as well. *GTA IV* replaced the zippily insubstantial, slippery-tired cars and motorcycles of *Vice City* and *San Andreas* with vehicles of brutely heavy actuality. While the crashes in *Vice City* are a ball, *GTA IV*'s car crashes are sensorily traumatic, often sending a screaming Niko through the windshield and into oncoming traffic. Running over a few stick-legged rag dolls in *San Andreas* is always good for a nasty-minded laugh, but running people down in *GTA IV* often leaves your bumper and headlights smeared with blood—evidence that gruesomely carries over into in-game cut scenes—and the potato-sack thud with which pedestrians carom off your windshield is, the first time you hear it, deeply disturbing. Liberty City is also more eccentrically populated. While there are plenty of people walking about Vice City and San Andreas, the character models are repetitive. I came across the latter's shirtless, red-Kangol-wearing LL Cool J doppelgänger so many times I started shooting the dude out of general principle. Liberty City's citizens have far more visual and behavioral distinction. (The bits of dialogue you overhear while walking down the street are some of the game's funniest: "You know," one cop cheerfully admits to another, "I *love* to beat civilians.") Discovering who panics and who decides to stop and duke it out with you when you try to steal a car is one of the *GTA* games' endless fascinations. When a Liberty City guy in a suit unexpectedly pulls out a Glock and starts firing it at you, you are no longer playing a game but interacting with a tiny node of living unpredictability. The owner of one of the first vehicles I jacked in Liberty City tried to pull me out of the car, but I accelerated before she succeeded. She held on to the door handle for a few painful-looking moments before vanishing under my tires in a puff of bloody mist. With a nervous laugh I looked over at my girlfriend, who was watching me play. She was not laughing and, suddenly, neither was I.

GTA IV's biggest advance from its predecessors was the quality of its apocalyptic satire. (The *GTA* games are not made by Americans and probably could not be made by Americans. Volition's *Saints Row* series, the most popular American-made *GTA* imitator, all but proves this, offering a vision of American culture that is unlikably frat-boyish and frequently defensive.) *Vice City* and *San Andreas* are too often content, satire-wise, to amuse themselves with stupid puns. While *GTA IV* has its share of pun gags (a chain of Internet cafés called Tw@, a moped known as the Faggio, an in-game credit card called Fleeca), and a number of simply dumb gags (its Statue of Liberty holds not a torch but a coffee cup), many of the haymakers it swings at American excess and idiocy make devastating contact. Much of the best material can be heard while listening to commercials on one of the game's nineteen radio stations. An Olive Garden–ish restaurant chain known as Al Dente's promises "all the fat of real Italian food, with a lot less taste and nutrients!" Broker's emo station, Radio Broker, uses "The station hipsters go to to say they've heard it all!" as its call sign. WKTT, Liberty City's conservative talk radio station ("Because democracy is worth suppressing rights for"), has as its Rush Limbaugh one Richard Bastion, a man given to pronouncements such as "Knowing you're always right—*that* is real freedom" and "Sodomy is a sin—even if I *crave* it." One of my favorite things to do in Liberty City is to retire to Niko's apartment and watch television. A brilliant cartoon show called *Republican Space Rangers* offers one of the most Swiftian portrayals of George W. Bush's foreign policy to be found in any medium: The Rangers' spaceship is shaped like a giant phallus and guided by an "insurgescan"; after annihilating a planet the Rangers do not even deign to visit, they commend themselves for "freeing mankind." The game's spleen shows most splendidly with Weazel News, its barely exaggerated version of Fox News. One Weazel newscast opens: "In a bloody terrorist attack that will surprise nobody . . ." At the crime scene itself, the on-the-spot reporter tells us that it is "a madhouse! We've got policemen signing book deals and firemen holding hoses and being photographed for Christmas calendars!"

Is Liberty City a metaphor for New York City, an imitation of New York City, or an exaggeration of New York City? The strength of Liberty City—a carefully arranged series of visual riffs on how New York City *looks* and *feels* rather than a street-by-street replication—is that, almost instantly, it becomes itself. As you learn Liberty City's streets and shortcuts, you are reminded of various real-life places—the cobbled streets of the Meatpacking District in the Meat Quarter, the shadowy concrete canyons of Midtown in Star Junction, the long weedy avenues of the South Bronx in South Bohan, the sterile pleasantness of Battery Park City in Castle Garden City—but these approximations quickly molt their interest. Soon you are thinking, *Oh, I need a new car, and can steal one from that Auto Exotica dealership right around the corner from here,* or *I can pick up Molotov cocktails near that Firefly Island bowling alley,* or *If I call Little Jacob right now, he will meet me in that alley by Star Junction Square, but if I call him two blocks from here, I'll have to find him underneath the East Borough Bridge.* I lived in New York City for close to a decade but have never played *GTA IV* while living there. To my delight, I found that *GTA IV* made me less homesick for the city. For me, Liberty City is an aggregate of surrogate landmarks and memories and the best way I have—short of reading a novel by Richard Price (whose *Lush Life* was one of the two novels I actually finished in the last year)—to remind me of what I love about the city it mimics.

To anyone who has not played the *GTA* games, this may be hard to swallow. What many without direct experience with the games do know is that they allow you to kill police officers. This is true. *GTA* games also allow you to kill everyone else. It is sometimes assumed that you somehow get points for killing police officers. Of course, you do not get "points" for anything in *GTA IV.* You get money for completing missions, a number of which are, yes, monstrously violent. While the passersby and pedestrians you slay out of mission will occasionally drop money, it would be hard to argue that the game rewards you for indiscriminate slaughter. People never drop that much money, for one, and the best way to attract the attention of the police, and begin a hair-raising transborough chase, is to hurt an innocent person. As for the infamous cultural trope that in *GTA* you can hire a prostitute, pay her, kill her, and take her money, this is also true. But you do not have to do this. The game certainly does not *ask* you to do this. Indeed, after being serviced by a prostitute, Niko will often say something like, "Strange. All that effort to feel this empty." Outside of the inarguably violent missions, it is not what *GTA IV* asks you to do that is so morally alarming. It is what it allows you to do.

GTA IV does have ideas about morality, some of which are very traditional. Many of the game's least pleasant characters are coke addicts, for instance. Niko is never shown imbibing any illegal substance, and when he gets drunk and plants himself behind the wheel of a car, the dizzily awhirl in-game camera provides an excellent illustration of why drunk driving is such a prodigiously bad idea. Finally, one surprisingly affecting mission involves Niko having to defend his homosexual friend Bernie from some thuggish gay-bashers in Middle Park. These are something less than the handholds of moral depravity.

Indeed, one criticism of GTA IV has to do with its traditional morality. Niko is shown during framed-narrative cut scenes to struggle with being asked to do such violent things, but while on furlough from these cut scenes Niko is able to behave as violently as the gamer wishes: Ludonarrative dissonance strikes again. But the game *does* attempt to address this. When Niko hits an innocent person in a car, he often calls out, "Sorry about that!" A small concession to acknowledging Niko's tormented nature, perhaps, but an important one. (Neither Tommy nor C.J. ever shows such remorse.) I chose to deal with GTA IV's ludonarrative dissonance in my own way. While moving through the gameworld, I did my best not to hurt innocent people. There *was* no ludonarrative dissonance for me, in other words, because I attempted to honor the Niko of the framed narrative when my control of him was restored.

There is no question, though, that GTA IV's violence can be extremely disturbing because it feels unprecedentedly distinct from how, say, films deal with violence. Think of the scene in *Goodfellas* in which Henry, Tommy, and Jimmy kick to death Billy Batts in Henry's restaurant. Afterward, they decide to put Batts's body in the trunk of Henry's car and bury it the forest. Of course, Batts is not yet dead and spends much of the ride to his place of interment weakly banging the trunk's interior. When Batts is discovered to be alive he is repeatedly, nightmarishly stabbed. The viewer of *Goodfellas* is implicated in the fate of Billy Batts in any number of ways. Most of us presumably feel closest to Henry, who has the least to do with the crime but is absolutely an accomplice to it. Henry's point of view is our implied own. Thus, we/Henry, unlike Tommy and Jimmy, retain our capacity for horror. Henry's experience within that horror is the scene's aesthetic and moral perimeter. In *GTA IV,* Niko is charged with disposing of the bodies of two men whose deaths he is partially responsible for. You/Niko drive across Liberty City with these bodies in the trunk to a corrupt physician who plans to sell the organs on the black market. Here, the horror of the situation is refracted in an entirely different manner, which allows the understanding that *GTA IV* is an engine of a far more intimate process of implication. While on his foul errand, Niko must cope with lifelike traffic, police harassment, red lights, pedestrians, and a poorly handling loaner car. Literally thousands of in-game variables complicate what you are trying to do. The *Goodfellas* scene is an observed experience bound up in one's own moral perception. The *GTA IV* mission is a procedural event in which one's moral perception of the (admittedly, much sillier) situation is scrambled by myriad other distractions. It turns narrative into an active experience, which film is simply unable to do in the same way. And it is moments like this that remind me why I love video games and what they give me that nothing else can.

An alkaloid drawn from a South American shrub, cocaine has been used by human beings for at least a thousand years and spectacularly abused for quite a bit less than that. Its familiar form as a white powder is yet another product of Teutonic ingenuity, for it was a German scientist who isolated the fun, psychoactive part of cocaine. An Austrian named Freud was among the first to study it seriously. (His initial findings: Cocaine was *terrific.*) Until 1914, cocaine could be legally purchased in US drugstores, parlors, and saloons, and was most often prescribed by doctors as a cure for hay fever. "Cocaine," Robert Sabbag tells us in the smuggling classic *Snowblind,* "has no edge. It is strictly a motor drug. It does not alter your perception; it will not even wire you up like the amphetamines. No pictures, no time/space warping, no danger, no fun, no edge. Any individual serious about his chemicals—a heavy hitter—would sooner take thirty No-Doz. Coke is to acid what jazz is to rock. You have to appreciate it. *It does not come to you.*"

Cocaine has its reputation as aggression unleaded largely because many who are attracted to it are themselves aggressive personalities, the reasons for which are as cultural as they are financial. What cocaine does is italicize personality traits, not script new ones. In my case, cocaine did not heighten my aggression in the least. What it did, at least at first, was exaggerate my natural curiosity and need for emotional affection. While on cocaine, I became as harmlessly ravenous as Cookie Monster.

This stage, lamentably and predictably, did not last long. Doing cocaine for more than a couple of days is a little like falling in love with someone who is attractive, friendly, adoring, devious, manipulative, evil, and congenitally incapable of loving you in return. But this person feels so unnaturally good, and makes you feel so unnaturally good about yourself, that you accept this as a fair bargain. When the deal you make with cocaine sours—and it will—your mind is as empty as a pasture, your basal ganglia shredded. You are now the moon to cocaine's sun: With it, you are bright indeed; without it, you are nothing more than a cratered rock stupidly afloat in space. You want to glow again. You do more cocaine. You do not glow—but you do feel somewhat normal. Soon you are doing cocaine not to feel radiant but to feel like yourself. Cocaine is no longer a sun but a hangman; this is how his noose tightens. And around my neck the rope tightened more quickly than I could have imagined.

A large portion of my last two months in Las Vegas was spent doing cocaine and playing video games—usually *Grand Theft Auto IV.* When I left Vegas, I thought I was leaving behind not only video games but cocaine. During the last walk I took through the city, in May 2008, I imagined the day's heat as the whoosh of a bullet that, through some oversight of fate, I had managed to dodge. (I was on cocaine at the time.) Even though one of the first things I did when I arrived in Tallinn was buy yet another Xbox 360, I had every intention to obey one of my few prime directives: rigorous adherence to all foreign drug laws. I had been in Tallinn for five months when, in a club, I found myself chatting with someone who was obviously lit. When I gently indicated my awareness of this person's altered state, the result was a magnanimous offer to share. Within no time at all I was back in my apartment, high on cocaine, and firing up my Xbox 360. By the week's end, I had a new friend, a new telephone number, and a reignited habit. I played through *Grand Theft Auto IV* again, and again after that. The game was faster and more beautiful while I was on cocaine, and breaking laws seemed even more seductive. Niko and I were outlaws, alone as all outlaws are alone, but deludedly content with our freedom and our power. Soon I was sleeping in my clothes. Soon my hair was stiff and fragrantly unclean. Soon I was doing lines before my Estonian class, staying up for days, curating prodigious nosebleeds, and spontaneously vomiting from exhaustion. Soon my pillowcases bore rusty coins of nasal drippage. Soon the only thing I could smell was something like the inside of an empty bottle of prescription medicine. Soon my biweekly phone call to my cocaine dealer was a weekly phone call. Soon I was walking into the night, handing hundreds of dollars in cash to a Russian man whose name I did not even know, waiting in alleys for him to come back—which he always did, though I never fully expected him to—and retreating home, to my Xbox, to *GTA IV*, to the electrifying solitude of my mind at play in an anarchic digital world. Soon I began to wonder why the only thing I seemed to like to do while on cocaine was play video games. And soon I realized what video games have in common with cocaine: Video games, you see, have no edge. You have to appreciate them. *They* do not come to *you*.

The world of *GTA IV* is not as open as it initially seems. The number of buildings you can enter is negligible; those few you can rarely provide anything to do other than walk in, look around, and maybe steal the cash from the register. *GTA IV's* mini-games—darts, bowling, billiards, strip-club lap dances—are uninteresting, and one sorely misses the taxi and ambulance driver mini-games of *Vice City* and *San Andreas* (which if nothing else provided outlets for socially beneficial behavior). Liberty City's comedy club has in rotation several rather good five-minute stand-up bits by Katt Williams and Ricky Gervais, and its television and radio stations are always entertaining, but these are not very gamelike activities. Rather, they are examples of traditional entertainment that happen to be embedded in a video game, though they are no less commendable for that. Once you have played *GTA IV* long enough, it occurs to you that, as real as Liberty City seems, you have no hope of even figuratively living within it. Accident or no, this is thematically coherent: Niko is a newcomer to and outsider in Liberty City, much of which is as fictionally inaccessible to him as it is literally inaccessible to us.

Although Niko has a cell phone, and an ever-fattening docket of friends to call, only a few can be rung up out of mission and asked out on "dates." You can then go play darts, bowl, or play billiards; visit the comedy club, strip club, or cabaret club; drink in a bar or go get something to eat at a surprisingly limited number of establishments. None of these activities are taken up because they are fun. They are taken up, rather, to win the influence of your friend or date, and I frequently wondered why such a prominent part of the game was handled in such a repetitive manner and supported by such a dearth of options.

Almost all of my fondest memories of *GTA IV* are anecdotal. The time I sniped the pilot of a zooming-by news chopper while standing on the GetaLife (read: MetLife) building and watched it whirlingly plunge down into the street and explode. The time a collision launched me from my motorbike and sent me sailing harmlessly through the girders of the Algonquin Bridge and into the East River hundreds of feet below. The time I used a few errantly parked city buses and garbage trucks to create a massive traffic jam in Star Junction Square, dropped a single grenade, and ran like hell as the cars blew up, one after another, for what felt like minutes. (The really violent stuff I did in games whose progress I did not save, so as to preserve my Niko's moral integrity.) The wonderful thing about the earlier *GTA* games was that they allowed anecdotally arresting things to occur while engaged in an otherwise scripted mission. *GTA IV* selectively, and thus frustratingly, abandons this idea. Some scripted chase-fights offer enemies who are inexplicably immune to damage until they reach a certain point on the game map. This is a problem because the game gives you no inkling as to which kind of enemy you are facing. Some narratively important chase-fights are not regulated in this way and some narratively unimportant chase-fights are. In one (important) chase, you have to swerve around a garbage truck that abruptly pulls out in front of you. Exciting the first time, frustrating the third, boring the fifth—and the game forces you to avoid the garbage truck because the enemy and his car cannot be damaged until after he passes it. In another (unimportant) chase-fight, you have to pursue two men on motorcycles through Liberty City's subway system. The first biker can be taken down quickly but the second refuses to take any damage, no matter how many times you shoot him, until you have dodged enough oncoming subway cars. The first time I played *GTA IV* I thought this mission was one of the most amazing I had ever experienced. When I realized that the first biker you damage becomes vulnerable, thereby making the other invulnerable, the once-thrilling chase seemed contaminated, arbitrary. Missions such as this nullify gamer skill and creativity because they force him to experience scripted events in an unalterable way, which goes against the whole spirit of what made earlier *GTA* games so revelatory.

The least interesting parts of the game are those that show the strongest authorial hand—and yet the part of *GTA IV* that affected me most is authored with an unopposed authorial hand, which brings me up short from being able to say with confidence that games are affecting *because* of gamer agency. This scene occurs near the end of the game, when Niko comes face-to-face with the man who betrayed their unit back in the Balkans, and who is now a pathetic, drooling, sore-covered, drug-addicted wretch. It would be pointless to describe the scene in much detail, but I will say that it is so well acted, written, and staged that it would not be out of place in any violent masterpiece, whether filmic or literary. What gives the scene its power is Niko's imploded recognition of his own moral ruin when he learns why this man betrayed him and his friends, which Niko had obviously imagined as an act of wicked grandeur. "You killed my friends for a thousand dollars?" Niko asks quietly, his voice breaking. Every time I have watched this scene, no matter how hard I fight it, tears fill my eyes when Niko's voice cracks, and they did again, just now, while thinking about it. When the scene concludes, you have your choice: kill the traitor or walk away. I struggled with my decision, and it feels almost too personally revealing to share what I did my first time through *GTA IV.* (I will share what I did my second time through: I walked away, hopped in a nearby semi, and ran the man over repeatedly.) Until this point, the radio has been your great companion—you have fishtailed into flocks of pedestrians to MC Lyte's "Cha Cha Cha," evaded SWAT teams to Philip Glass's "Pruit Igoe," and enjoyed hooker tug jobs to R. Kelly's "Bump N' Grind," sometimes scratching your head at the music–moment dissonance and sometimes winning the equivalent of a music–moment lottery—but while driving away from the aftermath of his decision, Niko, for the first and only time in the game, turns the radio off and tells his cousin, Roman, to stop talking. The wound this scene has left is too dirty to sterilize with anything other than silence. On the long ride home, Niko has only your thoughts to accompany him.

There are times when I think *GTA IV* is the most colossal creative achievement of the last twenty-five years, times when I think of it as an unsurpassable example of what games can do, and times when I think of it as misguided and a failure. No matter what I think about *GTA IV,* or however I am currently regarding it, my throat gets a little drier, my head a little heavier, and I know I am also thinking about cocaine.

Video games and cocaine feed on my impulsiveness, reinforce my love of solitude, and make me feel good and bad in equal measure. The crucial difference is that I believe in what video games want to give me, while the bequest of cocaine is one I loathe and distrust. As for *GTA IV,* there is surely a reason it was the game I most enjoyed playing on coke, constantly promising myself "Just one more mission" after a few fat lines. (In Vegas and Tallinn, "One more mission" became the closest thing I have ever had to a mantra.) For every moment of transcendence there is a moment in the gutter. For all its emotional violence there are long periods of quiet and calm. Something bombardingly strange or new is always happening. You constantly find things, constantly learn things, constantly see things you could not have imagined. When you are away from it, you long for its dark and arrowy energies. But am I talking about video games or cocaine? I do know that video games have enriched my life. Of that I have no doubt. They have also done damage to my life. Of *that* I have no doubt. I let this happen, of course; I even helped the process along. As for cocaine, it has been a long time since I last did it, but not as long as I would like.

So what have games given me? Experiences. Not surrogate experiences, but actual experiences, many of which are as important to me as any real memories. Once I wanted games to show me things I could not see in any other medium. Then I wanted games to tell me a story in a way no other medium can. Then I wanted games to redeem something absent in myself. Then I wanted a game experience that points not toward but *at* something. Playing *GTA IV* on coke for weeks and then months at a time, I learned that maybe all a game can do is point at the person who is playing it, and maybe this has to be enough.

I still have an occasional thought about Niko. When I last left him he was trying to find all the super jumps hidden around Liberty City, which is a strange thing for a wanted fugitive to be doing. I know he is still there, in his dingy South Bohan apartment (my Niko is definitely a South Bohan kind of guy. That penthouse near Middle Park? I never let him near the place), waiting for me to rejoin him. In early 2009 Rockstar released some new downloadable content for *GTA IV, The Lost and Damned*, in which you follow the narrative path of Johnny Klebitz, an incidental character in Niko's story (his most memorable line: "Nothing like selling a little dope to let you know you're alive!") but whose story, it turns out, intersects with Niko's in interesting ways. I played this new *GTA IV* story for a few hours but gradually lost interest and finally gave up. I realized, dismayingly, that a lot of what powered me through *GTA IV* had been the cocaine, though it is still my favorite game and probably always will be. I was no longer the person I had been when I loved *GTA IV* the most and, without Niko, Liberty City was not the same.

Niko was not my friend, but I felt for him, deeply. He was clearly having a hard go of it and did not always understand why. He was in a new place that did not make a lot of sense. He was trying, he was doing his best, but he was falling into habits and ways of being that did not reflect his best self. By the end of his long journey, Niko and I had been through a lot together.

APPENDIX: AN INTERVIEW WITH SIR PETER MOLYNEUX

Anyone who plays video games will probably have a list of titles that he or she wishes I had talked about in this book. As it happens, I myself have such a list. Games I did discuss but wound up cutting include *Shadow of the Colossus, Half-Life 2,* and *Assassin's Creed,* while games I intended to discuss but never found a way to include *Indigo Prophecy, Ico, Perfect Dark, Mirror's Edge,* and *Eternal Sonata.* (This is to say nothing of some wonderful games I have played since finishing the book, including EA's *Dead Space* and Naughty Dog's *Uncharted: Drake's Fortune.**) Two of the games I was most eager to discuss before I began this book were Hideo Kojima's *Metal Gear Solid 4: Guns of the Patriots* and Lionhead's *Fable II,* both of which, to my regret, turn up in *Extra Lives* only in passing.

---

* And this is to say nothing of some wonderful games I've played since inserting this note etc.

I generated many pages of notes and observations about both games and spent two very enjoyable evenings with the video-game critic Leigh Alexander—the Western world's resident Hideo Kojima expert—playing *Metal Gear Solid 4,* which manages to be as graphically beautiful and mechanically complex as any game ever and, at the same time, somehow deliberately backward-looking aesthetically (not to mention its many mescaline-grade weirdnesses, which include a smoking monkey in a silver lamé diaper). Alexander's take on why *MGS 4* is this way is so interpre-tively brilliant that, as she spun it out for me, my skeptical frown gave way to a dropped jaw and many thoroughly persuaded nods. Unfortunately, as Alexander admits, the story of the *Metal Gear Solid* games is "incomprehensible" to anyone not deeply steeped in its lore, and trying to summarize that story here would be akin to a one-page encapsulation of *War and Peace.* The following, then, taken from my interview with Alexander, is for *Metal Gear Solid* brown belts and above:

ALEXANDER: I don't see the game as being solely metaphorical but I think there's an intended subtext, which is the journey of the game designer whose methodology is out of date. After *Metal Gear Solid 3,* Kojima said, "I don't want to make *Metal Gear* games any-more." But here was this new PS3, and it looked like it might allow Kojima to execute his vision to the fullest. Remember, Kojima is a national hero in Japan, and Sony, a Japanese company, approached him and said, "Do you really want to stop when you could make the ultimate stealth game on this piece of ultimate hardware?" So here's Snake, a man who doesn't believe he's a hero, with one more job to do, and technology is what's going to make it possible. But the promises of technology are always inhuman and dis-appointing—and Kojima has pretty much said that the PS3 did not live up to what he was promised it could do. In *Metal Gear*

Solid 4, additionally, Snake is old. The player is very deliberately made to feel sympathy for this guy who used to be so strong and unstoppable and is now just a relic. The cinematography of the game—whether or not you hate the epic cut scenes—creates a ridiculous amount of empathy for this old guy being constantly eclipsed by younger, faster guys, like Johnny. By ending up with Meryl at the end of the game, Johnny begins to visually resemble the young Snake—even down to the mullet! The subtext is obvious: Meryl likes Johnny because he reminds her of a young Snake. I believe that this is Kojima's concession to having been eclipsed by Western game developers. You have this young, dumb, blond guy who used to be a fuckup, and he's the one who gets the girl. What is the most interesting thing about Johnny? He had not been corrupted by the promises of new technology. He was dumb, but he was pure. So Kojima is taking this buffoon and saying, "Man, the stupid white kid knew better all along, and now he's taken over." The war in the game is the *console* war.

Because I could not say this better, and because I admire *Metal Gear Solid 4* more than I enjoy playing it, I found I had no way *into* discussing its gameplay other than by cribbing Alexander's extremely persuasive analysis of what it means.

I had a related problem with *Fable II,* another game I deeply and genuinely admire. When the time came for me to write about it, however, I froze. I could never find a solid place from which to explore a game for which I *mostly* felt admiration. This was especially disappointing because *Fable II's* legendary designer, Sir Peter Molyneux, was kind enough to grant me an interview at the 2009 Game Developers Conference in San Francisco.

Fable II is an open-world fantasy RPG that allows you to quest, pose for sculptures, get married, have children, get *gay* married, cheat on your spouse, use condoms, get sexually transmitted diseases, get fat, slim down, own a dog, find treasure, buy houses, teach your dog tricks, gamble, work as a bartender, fight, learn spells, pay bards to sing epic songs of your exploits, chop wood, decorate your house, save the world, and kill a friend. *Fable II*'s refusal to traffic in video-game clichés (its final boss fight is one of the most swiftly and unexpectedly resolved in game history), its mischievousness (rarely has any game with a "bad–good" behavior mechanic made being bad so guiltlessly fun), and its sense of humor ("Why," one aristocratic woman said when my female character sexually propositioned her, "I haven't done that sort of thing since my dormitory days!") make it, without question, a game of rarefied formal sophistication—a strange claim to make for a game that uses a cartoony "expression wheel" as its character-to-character interface. In short, when you want to "talk" to someone in *Fable II,* you hold down a button, bring up the expression wheel, select which "emotion" you would like to communicate (happiness, aggression, playfulness, amorousness), and then select a distinct expression of that emotion (laughing, muscle-flexing, farting, bedroom eyes). Communication in *Fable II* is thus largely gestural, the audacity of which is especially daring when one considers the difficulty video games have had with using gesture as a meaningful element of the game experience.

What held me back from finally loving *Fable II* in the heedless way I love other, less admirable games, I am not certain. I went into my interview with Molyneux—one of the nicest and most intelligent people I met while researching this book—hoping, in part, to get an answer. I am not sure I got one. But I did discover why Molyneux's reputation as one of the few undisputed geniuses of game design undersells him if it does anything at all.

What follows is a lightly abridged transcript of our conversation.

TCB: Down at the expo hall this morning I was playing *Resident Evil 5* and thinking a lot about the benefits and deficits of photorealistic representation—that is, the problems and solutions photorealism creates for games—and I realized that one of the many things that bugs me about *Resident Evil 5* is that the quality of the representation graphically is inconsistent with the cartoonish results you get when you're shooting people, which is what the whole game is based around. Enemies just go flying like Looney Tunes characters. Then I thought of *Fable II*, which is, representationally, a realistic game—storybook realism, I would say—but which also has this wonderfully unrealistic expression wheel. Somehow, though, the gamer never senses this same kind of dissonance. Could you speak to how you walked that line?

MOLYNEUX: When we first started with the *Fable* franchise we looked around for a visual style that wouldn't be too exact. It's not just the pixels on the screen; as you say, it's the animation, it's the speech, it's the timing, it's the fighting—all of those things have to come together. We're very close to realism, but the closer we are, the further we are away, weirdly enough. So the visual style we picked for the first *Fable* was Tim Burton's *Sleepy Hollow,* which had that same kind of mixture. It's almost abstract; the colors are a little bit brighter. I think that, subconsciously, that keeps you from thinking, *Hey, that person's eyebrows are not moving in the right way.* When we came to *Fable II,* we looked around for a bit of a change in that visual style, but not to go too far over that line. We chose a film called *Brotherhood of the Wolf.* Again, you look at it and you know that maybe this never was a place you could go to or visit, but it was close enough to reality that you weren't estranged by it.

TCB: So would you agree with the idea that when realism is the goal, it also becomes the problem?

MOLYNEUX: Absolutely, it is. To really achieve realism, what you're dealing with, when it comes down to it, is something called neurolinguistic programming. There are hundreds of thousands of tiny little messages that our brains are picking out from faces, the environment, the lighting, the time of day, the amount of dust in the atmosphere, which gives us, at best, a *sense* of reality. And whilst we're making strides to achieve that in games—and I have no doubt we will achieve that—we are still a ways off. Some people call it the Uncanny Valley. I don't think the Uncanny Valley exists *if you choose the right stage.* If you chose the wrong stage . . . it's like trying to cast a Shakespearean play with cats. It doesn't work. One of the things you've got to remember is that games are made by people who are, first, computer-game developers. It has taken the film industry and the television industry and the theater decades and decades and decades to get some principles right. The problem we had with *Fable II,* and it is a problem a lot of games have, is that when you come to the "story," you have to wait, because there's all this technology that's being created. You have to create your scripting engine, you have to create your environments, you have to create your gameplay, you have to create your controls . . . you're going away all along, and all of that stuff is not finished until, probably, two months before the end. Well, guess what? That's when you've got to start editing your story, and that's just not enough time.

TCB: It's a weird process.

MOLYNEUX: It's a *very* weird process. It's kind of like trying to shoot a film and spending 90 percent of your time making the set and 10 percent of the time shooting the actors. In film they shoot a huge amount of footage and edit that down. In *Fable II,* what we did was, first, realize that we were really *rubbish* at telling stories.

Then we found this director who was willing to actually talk to us about staging. And that was: You have a script. You have some actors. You figure out how to position them and what their gestures are going to be and how they will behave and where certain things are going to happen. We hired a soundstage, a place called Shepperton, and went into a huge white room, put all the actors in there, gave them all the script, sat back, and watched this director *direct* all this stuff. And, my god, it was just an amazing moment when we realized that the nuances we were trying to communicate, the emotion we were trying to get into our characters, was driven *solely* and *purely* by dialogue. And a lot of what we had written would have worked much better on the radio than it would have on the screen. This director would say, "Right, let's have this person walk here, and let's ask this actor, 'What would you do if you just heard this piece of news?' " Watching the actors improvise and get into the characters was an incredible experience. We did that for the entirety of the story, so that we could *feel* what the story was like before it was implemented in the game. What we discovered—which was quite amazing and is so true about a lot of video games—is that the story we had written was so wordy, and so slow-paced and turgid, that a lot of the dialogue we could rip out. We already *had* a lot of the emotion. We didn't have to ram it down the audience's throat. As human beings, we're used to getting a *feel* of what people are thinking by their gestures, but now we're using that technique in games and in a position to achieve something very special.

TCB: The use of the expression wheel is a way around needing huge amounts of dialogue, then?

MOLYNEUX: The expression wheel is a way for the player to emote, but I'm not terribly happy with it. There's an enormous amount more that could be done with it. The way it works is that it cuts down on the necessity of dialogue, yes, but the way it didn't work was when it did not provide the right emotional connection and came across as a little bit trivial. The great thing about it is the stories people were able to make up in their own minds about what they were doing. I'll give you an example. There was this journalist that came in about two hours ago who was talking about something that happened to him and his wife while they were playing *Fable II,* and it was all to do with the child they had had in the game. They went off adventuring and came back to see their son for the first time. And so they thought, "What expression should we do for this child?" This child was saying, "Mommy, Mommy—you're home! Where have you been?" So they decided to make the little kid laugh and tried to do the "sock puppet" expression. But they messed it up in doing it, picked the wrong expression, and ended up punching the air. Their kid got really scared and said, "Mommy, don't hurt me!" That moment became unbelievably emotional for them. And that, I think, is where the expression wheel worked, because it allowed people to make their own stories up without it being totally encapsulated by what I wanted to do. And that is an amazing place for us to get to.

TCB: I can tell you that when I played *Fable II* I became a slutty lesbian bigamist who had tons of children, all of whom I abandoned.

MOLYNEUX: That's fantastic!

TCB: I have to say, *Fable II* probably made me laugh more than any other game.

MOLYNEUX: Oh, thank you.

TCB: At the beginning of the game, for instance, I was breaking all the crates and wondering why I wasn't finding anything in them. Then that one load screen comes up and says, "Breaking crates is good fun, but you don't think someone would actually *hide* anything in one, do you?" When I read that, and laughed, I wondered, Is *comedy* the great untapped game genre?

MOLYNEUX: I remember that crate moment; it's a funny thing you should mention it. There's a lot of debate and talking and doubt when you create a game, and that crate moment is very interesting. I can remember us saying, "Well, we want to put loads of stuff in the crates." And I was saying, "*Why* do we want to put stuff in crates? We all know there's no stuff in crates. Are you really going to ask people to go 'round breaking every single crate? That's not a game. That's tedium. Let's just make people laugh with that one sentence." I think it's fantastic that that worked. And now I've forgotten your question.

TCB: Is comedy the untapped game genre.

MOLYNEUX: I think it's so rare to make people laugh in any form of entertainment. When you're in a pub, and you play that game, "What's your favorite film?"—it's easy to say what your favorite horror film is, what your favorite action film is, but your favorite comedy? That's tough.

TCB: Are you a Monty Python fan, by any chance?

MOLYNEUX: Monty Python is fantastic. You can see the influence of Python on *Fable*.

TCB: Yes.

MOLYNEUX: The best humor is the humor you come and discover, like the crate joke. I hope, if there are any ideals we stick to in the *Fable* universe, I hope that humor is one of them. But comedy is very, very hard to do. We've got someone called Mark Hill who is absolutely, blindingly good. He's responsible for an awful lot of the dialogue. Someone else named Richard Bryant, who's actually an American writer, is responsible for a lot of it as well. This isn't something directed by me. I don't say, "Okay, we're at Funny Level Fifteen, let's take it to Seventeen." It's something that comes very naturally to those guys.

TCB: I know a lot of people talk about *Fable* and other games as having "moral choices," but what I liked about *Fable II* was that it seemed more interested in questions surrounding matters of moral choice rather than the specific moral choices themselves. The game encourages you to be bad, doesn't it?

MOLYNEUX: It *tempts* you to be bad.

TCB: Okay. So you would say—

MOLYNEUX: *That's* the theme. The *temptation* to be bad. Originally, when I first came up with the idea of doing a game that lets you be good or evil, I expected everyone to be evil. But the reverse is true. It's quite fascinating how it is *very* country-specific—the percentage of people who are good and bad. Americans, fascinatingly, have the highest percentage of good guys.

TCB: Really?

MOLYNEUX: I would have thought the opposite. A slightly constrained society, the American Dream, and all that. I would have thought there'd be some rebels.

TCB: We actually believe our own delusions.

MOLYNEUX: Yeah, I know! When we delved into it deeper, we asked a lot of psychologists why these trends were, and the theory was: Although you guys do have this American Dream, Americans feel more constrained by the thought of, "Well, there's no way I can even *tempt* myself by being evil. That would be really, really bad." Whereas people like the English are much more willing to play the multiple mass murderer and lesbian bigamist.

TCB: I used to very reliably play the "good" path in games and then go back and play the "bad" path. But now my play style is erratic, because I'm more interested in how games *respond* to these choices.

MOLYNEUX: I think that was, a little bit, one of the failures of *Fable II*. You kind of felt like you had to go back and play it again—and it's never going to work on the second play-through. You're never going to enjoy it as much. It's actually going to muddy those memories you've got of the game and the story if you play through it again.

TCB: I have to say, the one part where I couldn't do the "bad" thing was during the profoundly troubling Tattered Spire sequence, where you have the choice to torture people and put your friend out of his misery. I just could not do either thing. And I was so happy you put that stuff in there.

MOLYNEUX: That was going to be a lot, lot stronger, but it had to be weakened down for all the obvious political reasons. There's always this thought that, "Hey, the good guy never caves under torture, never caves under pressure," and I really wanted to push you and test you on that, and I really wanted you to feel like you were sacrificing something there.

TCB: How real were the consequences for not killing your friend? I mean, I know I lost permanent experience points by not killing him, but—

MOLYNEUX: You lost experience points, but we should have taken more experience, actually. We chickened out there. It was quite harrowing at one point. You had someone who was strapped to this machine and you were going to be asked to torture him, using these different devices. I wanted to get you to say, "No, I can't bring myself to do that."

TCB: Was cutting that out an internal decision or an external decision?

MOLYNEUX: This was the time when the world was in that topsy place where torture was even more politically sensitive, so we cut that out. The one that I found the most interesting, probably, if we'd done it a little better, was the bit where you had to beg for mercy when the Commandant was saying, "Beg! Beg again!" I think we could have done more with that, because it was just words. It wasn't beating or not beating people. It had the potential to be a lot more powerful. You watch film after film where good guys never beg for mercy. But how far do I have to push you before you beg for mercy?

TCB: For the record, I begged right away.

MOLYNEUX: And so you realized that what you had lost there was a little bit of your own self-respect.

TCB: Can we talk a bit more about the Tattered Spire sequence? I don't tend to read very much about games before I play them, because I want them to be fresh, so the Tattered Spire sequence I came to late at night, with no idea that it was coming. I see on the screen this amazing MANY YEARS LATER part title and suddenly I'm training to be an evil soldier in the Tattered Spire, with all my spells and weapons and clothing and items taken from me, and, my god, my *head* shaved—I mean, this is just not something you see in games. Ever. It was a total confounding of expectation. *Many years later?* What I loved about it was that it seemed—and this is going to sound a little pretentious—but it seemed a really brave aesthetic decision to have made.

MOLYNEUX: There were a lot of fights over that sequence.

TCB: I bet.

MOLYNEUX: But it felt like that.

TCB: Brave?

MOLYNEUX: A little. I wanted it to be even more emotional than it ended up being, because the whole point of it was upsetting the rhythm of the story. You had had all this success of finding Hammer and you had gone through the Arena, and you felt like a big, tough hero. And I wanted to strip everything away from you and say, "Hang on a second. You're not that big, you're not invincible, and every fight you face you aren't necessarily going to win." It worked to a certain extent, so it's fantastic to hear that you responded to it.

TCB: I loved it. My personal belief is that what makes works of art great is often what is weird and kind of flawed about them. And what I admire and appreciate about someone like Hideo Kojima is his eccentric insistence on forty-eight-minute-long cut scenes. I don't care if they don't "work." It seems like a personal vision. And that's what the Tattered Spire sequence felt like to me: a vision realized, convention and consequence be damned.

MOLYNEUX: I would love to talk to you about this other thing we're working on—but I can't—because you'll hear about this talk of "twenty-two minutes." And you're going to hear this soon. Why is it twenty-two minutes? It's something that happens in twenty-two minutes. It's not logical; it's something in my mind. When you see the announcement, you'll know what I'm talking about.

TCB: I won't press you—though I really, really want to. One of the people I've talked to while working on this is Jonathan Blow.

MOLYNEUX: Yes.

TCB: Do you agree with him that the forward progression of story and the "friction force" of challenge create structurally unsound narrative? That games can't tell stories in a certain sense because they're built on a flawed edifice?

MOLYNEUX: I don't know if I agree with that.

TCB: I don't know if I agree, either, but it's a very interesting argument.

MOLYNEUX: It is. You know, the thing about *Braid:* I loved it, I loved the atmosphere, I loved the visions, the softness of it. It kind of felt like a piece of silk you could run your hands through. It was a lovely, lovely game. But here's the thing that didn't work for me: It got so tough that my need and want to experience more of its world was absolutely challenged by my feeling that I wasn't clever enough. I hit this cerebral brick wall where I kept going back to find out more about the world, feeling more and more stupid. After a while, I thought, *This game is dumb.* Now I think I was wrong, by the way. But this was another fight we had in *Fable,* which is about the death mechanic.

TCB: *Fable II* has an unusually forgiving death mechanic. A lot of people accused the game of being too easy.

MOLYNEUX: If you're writing a game, why is it in so many games—even in games I've done—when the player dies, you ask the player to go back and reexperience what they've experienced before? Why do we do that? It just makes us feel stupid, and dumb, and we forget what the story is. We don't care about the characters anymore. Some guy is telling me what I need to do again, and I want to *kill* him if he tells me one more time! I think that's . . . well, thinking about story and narrative and gameplay, they should have a beat and rhythm that work together. You shouldn't have gameplay being this one big thing shouting, "I'm more important than you!" They should work together, in concert. And if they do, then what I really want you to feel is fantastic about the narrative and the rhythm of the story *and* feel fantastic as a player. It's what *you're* feeling, not what I'm feeling as a designer. That's what's important: what you're feeling.

TCB: Do you follow the indie game scene? A lot of the game writers I've met here seem to think that the indie game scene is the future.

MOLYNEUX: The funny thing is, we've been here as an industry before. Three years ago, these guys didn't exist. They weren't here. The entry level into the industry was so enormously high. If you asked me how to get into the game industry three years ago, I would have said, "Go to university, get a top degree, then go to work as a junior coder or designer and maybe in seven years' time you'll be a lead designer on a game." Now I can say to you, "Get a friend, smoke lots of dope, go in a room, come out when you've got a really good idea, and release it on Xbox Live Arcade." And you know what? That's where I was twenty years ago. I was one of those guys twenty years ago. I was doing a game called *Populous*. I was in a room. I had no idea what I was doing. I didn't know, really, anything about game design, or much of anything about programming, and I sort of came up with this concept. So yes, some of those people are going to be the future, but I don't think you can look even at *Braid* and think, *This is the future of games.* It's just one aspect of it.

TCB: I went to the Hideo Kojima lecture this morning, and he showed slides from the first *Metal Gear* game and then the most recent, and seeing those images in such close proximity made me realize, "My god—we've gone from petroglyphic rock art to the Sistine Chapel in twenty years!"

MOLYNEUX: I'm going to sell this hard, because I love what I do and I love this industry. Here's what's even more amazing: If I were to draw on the wall what a computer-game character was just twenty years ago it would be made up of sixteen-by-sixteen dots, and that's it. We've gone from that to daring to suggest we can represent the human face. And pretty much everything we've done, we've *invented.* There wasn't this technology pool that we pulled it out of. Ten, fifteen years ago, you couldn't walk into a bookshop and learn how to do it. There weren't any books on this stuff. They did not exist. Painting the ceiling of the Sistine Chapel? No. We had to invent architecture first. We had to quarry the stones. We had to invent the paint. That really is *amazing.* Think of word processors and spreadsheets and operating systems—they're all kind of the same as they were fifteen years ago. There is not another form of technology on this planet that has kept up with games. The game industry marches on in the way it does because it has this dream that, one day, it's going to be real. We're going to have real life. We're going to have real characters. We're going to have real drama. We're going to change the world and entertain in a way that nothing else ever has before.

ACKNOWLEDGMENTS

My first and biggest thanks to Cliff Bleszinski, Dave Nash, Lee Perry, Rod Fergusson, Chris Perna, Alan Willard, Ray Davis, and Tim Sweeney from Epic; Drew Karpyshyn and Heather Rabatich from BioWare; Clint Hocking and Cedric Orvoine from Ubisoft Montreal; John Hight from Sony Computer Entertainment; Sir Peter Molyneux from Lionhead; Joshua Ortega; and Jonathan Blow. Thank you to Debbie Chen and Joseph Olin, president of the Academy of Interactive Arts & Sciences, for speaking with me. Thank you, too, to the writers Chris Dahlen, Michael Abbott, Leigh Alexander, Geoff Keighley, Scott Jones, Rob Auten, Matthew S. Burns, Jamin Brophy-Warren, and Harry "the Media Assassin" Allen (who probably does not know that it was our conversation, years ago, at a Rockstar party, that first got me thinking about writing this book), for your work and the inspiration it frequently provided. A special thank-you to Heather Chaplin, who opened the door.

Thank you to Leo Carey and David Remnick at *The New Yorker* for allowing men with chainsaws to provide me entry into its pages. Thank you, as always, to Heather Schroder, Dan Frank, and Andrew Miller. Thank you to Oliver Broudy, for an early and important nudge, and Ross Simonini, for another. Thank you to Adrienne Miller, who read versions of these chapters many times. Thank you to Gary Sernovitz, the Skeptic. Thank you to Juliet Litman, for her excellent transcribing work. An especially huge and distended thank-you to Mark Van Lommel, whose enthusiasm, belief, and magic Rolodex in many ways allowed this book to be written.

To David Amsden (the finest sniper on Sera), Nathalie Chicha (my guitar hero), Jeff Alexander (Prophecy!), Dan Josefson ("My father? The president?"), Yrjö Ojasaar (who will never play as the Russians in *Civilization Revolution*), Jei Virunurm (the Force is strong with this one), Matthew McGough ("Oh, god! Help me!"), Kerle Kiik (Hendrix lives!), Jen Wang (Freebird), Joe Cameron (thanks for that neat pistol-whip trick, and a few others), Marc Johnson ("Pills here!"), Jason Coley (Master Chief), Gideon Lewis-Kraus (The Power of the Atom!), Paolo Bernagozzi (*il mio video fratello*), Hendrik Dey (Goooooooaaaaaaal!), Pierre-Yves Savard (zombicidal maniac), Nick Laird (shadow hide you), Arman Schwartz (Lego enthusiast), Juan Martinez ([undead groan]), and Owen King (Xbox 360 melter): You have been my most frequent video-game partners and opponents over the last few years; thank you for playing with me. To Maile Chapman, thank you for listening to the gestation of so many of these ideas, for being the first person to whom I showed *BioShock,* and for everything else. Thank you, finally, to Trisha Miller, who is, and ever will be, *my* extra life.